VOLUME ONE HUNDRED AND NINE

ADVANCES IN GENETICS

VOLUME ONE HUNDRED AND NINE

ADVANCES IN GENETICS

ELSEVIER

ACADEMIC PRESS

An imprint of Elsevier

Academic Press is an imprint of Elsevier
50 Hampshire Street, 5th Floor, Cambridge, MA 02139, United States
525 B Street, Suite 1650, San Diego, CA 92101, United States
The Boulevard, Langford Lane, Kidlington, Oxford OX5 1GB, United Kingdom
125 London Wall, London, EC2Y 5AS, United Kingdom

First edition 2022

ISBN: 978-0-443-18580-9
ISSN: 0065-2660

For information on all Academic Press publications
visit our website at https://www.elsevier.com/books-and-journals

Publisher: Zoe Kruze
Developmental Editor: Cindy Angelita Gardose
Production Project Manager: Abdulla Sait
Cover Designer: Greg Harris

Typeset by STRAIVE, India

Contents

Contributors

Gerald R. Smith
Division of Basic Sciences, Fred Hutchinson Cancer Center, Seattle, WA, United States

Suriyen Subramaniam
Interlake High School, Bellevue, WA, United States

Xuhua Xia
Department of Biology, University of Ottawa; Ottawa Institute of Systems Biology, Ottawa, Canada

CHAPTER ONE

RecBCD enzyme and Chi recombination hotspots as determinants of self vs. non-self: Myths and mechanisms

Suriyen Subramaniam[a,†] and Gerald R. Smith[b,*]
[a]Interlake High School, Bellevue, WA, United States
[b]Division of Basic Sciences, Fred Hutchinson Cancer Center, Seattle, WA, United States
*Corresponding author: e-mail address: gsmith@fhcrc.org

Contents

Abstract

Bacteria face a challenge when DNA enters their cells by transformation, mating, or phage infection. Should they treat this DNA as an invasive foreigner and destroy it, or consider it one of their own and potentially benefit from incorporating new genes

† Current address: University of California at Los Angeles, Los Angeles, CA, United States.

Advances in Genetics, Volume 109
ISSN 0065-2660
https://doi.org/10.1016/bs.adgen.2022.06.001

or alleles to gain useful functions? It is frequently stated that the short nucleotide sequence Chi (5′ GCTGGTGG 3′), a hotspot of homologous genetic recombination recognized by *Escherichia coli*'s RecBCD helicase-nuclease, allows *E. coli* to distinguish its DNA (self) from any other DNA (non-self) and to destroy non-self DNA, and that Chi is "over-represented" in the *E. coli* genome. We show here that these latter statements (dogmas) are not supported by available evidence. We note Chi's wide-spread occurrence and activity in distantly related bacterial species and phages. We illustrate multiple, highly non-random features of the genomes of *E. coli* and coliphage P1 that account for Chi's high frequency and genomic position, leading us to propose that P1 selects for Chi's enhancement of recombination, whereas *E. coli* selects for the preferred codons in Chi. We discuss other, previously described mechanisms for self vs. non-self determination involving RecBCD and for RecBCD's destruction of DNA that cannot recombine, whether foreign or domestic, with or without Chi.

1. Introduction

Potentially dangerous DNA frequently enters bacterial cells *via* various mechanisms. During transformation, naked DNA enters cells, in some species by specialized, high-efficiency up-take mechanisms, and can be integrated into the recipient genome by recombination. Phages actively inject their DNA into cells to initiate infection, which can kill the cells. In transduction, a rare phage particle contains bacterial DNA incorporated from the previous host's genome; the injected DNA, as in transformation, can either integrate into the recipient's genome or set in motion acts harmful to the recipient cell. Conjugative plasmids encode specialized machinery to transfer DNA directly into the recipient cell, thereby bypassing exposure of the DNA to a potentially destructive environment. In some cases, the plasmid also transfers bacterial chromosomal DNA along with the plasmid DNA.

In each case, the imported DNA can come from a species different from the recipient which, along with *E. coli* phages, we designate "foreign" or "non-self." DNA transfer can be directly observed in the laboratory, and these transfers likely occur in nature, too. It is inferred to occur frequently in nature from the presence, in unrelated species, of DNA "islands"—highly conserved DNA sequences with one or more genes, often functionally related indicating they travel together as a unit (Hacker & Kaper, 2000). Foreign, or non-self, DNA is more likely to be deleterious than self DNA, and it is reasonable that bacteria have ways to distinguish the two. They can profit from some DNA (self as well as beneficial non-self DNA such as islands encoding drug-resistance), but they are at risk of imminent death from other DNA (non-self, such as that of virulent phages).

How, then, do bacteria, such as *Escherichia coli*, distinguish self from non-self DNA? Numerous authors have stated that the short DNA sequence called Chi occurs at higher than expected frequency in the *E. coli* genome, that Chi is a signal allowing *E. coli* to tell self from non-self DNA, and that Chi's main function is to block degradation of DNA (e.g., Anderson & Kowalczykowski, 1998; Biswas, Maguin, Ehrlich, & Gruss, 1995; El Karoui, Biaudet, Schbath, & Gruss, 1999; Kobayashi, 1998; Kuzminov, 1995; Levy et al., 2015; Stahl, 2005; see Table S1 (Supplementary Material in the online version at https://doi.org/10.1016/bs.adgen.2022.06.001) for quotations). Chi (5' GCTGGTGG 3') is recognized by the *E. coli* RecBCD helicase-nuclease as it unwinds DNA rapidly [~1000 base-pairs (1 kbp)/s] and prompts RecBCD to nick the DNA at Chi and promote recombination between that DNA and other homologous double-stranded (ds) DNA in the cell (Smith, 2012) (Fig. 1; see discussion below for further explanation). RecBCD's activity can spur incorporation of incoming self DNA and aid evolution; it can also repair broken DNA in a cell with multiple copies of its chromosome (as typical in growing bacteria). But in other situations, RecBCD's nuclease activity can degrade DNA (Clark, Chamberlin, Boyce, & Howard-Flanders, 1966; Simmon & Lederberg, 1972). It is often stated that Chi instructs RecBCD to recombine self DNA (containing Chi) and to destroy non-self DNA (without Chi) (e.g., Cheng, Wilkinson, Chaban, & Wigley, 2020; Levy et al., 2015; Touzain, Petit, Schbath, & El Karoui, 2011). Below, we discuss and evaluate three widely-held views (dogmas) about RecBCD and Chi:

(1) In *E. coli*, Chi distinguishes self from non-self DNA.
(2) Chi occurs more frequently in *E. coli* DNA than expected.
(3) Chi converts RecBCD from a destructive nuclease into a DNA break-repair machine.

We then discuss mechanisms by which bacteria do discriminate between self and non-self DNA using RecBCD.

2. Chi is more frequent in many bacterial species and *E. coli* phages and plasmids than in *E. coli* itself

The view that Chi is the signal for *E. coli* to distinguish its own DNA from that of other species or phages implicitly assumes that Chi occurs more frequently in *E. coli* DNA than in other DNA. But this is not the case—the density of Chi is greater in many other bacterial species and *E. coli* phages

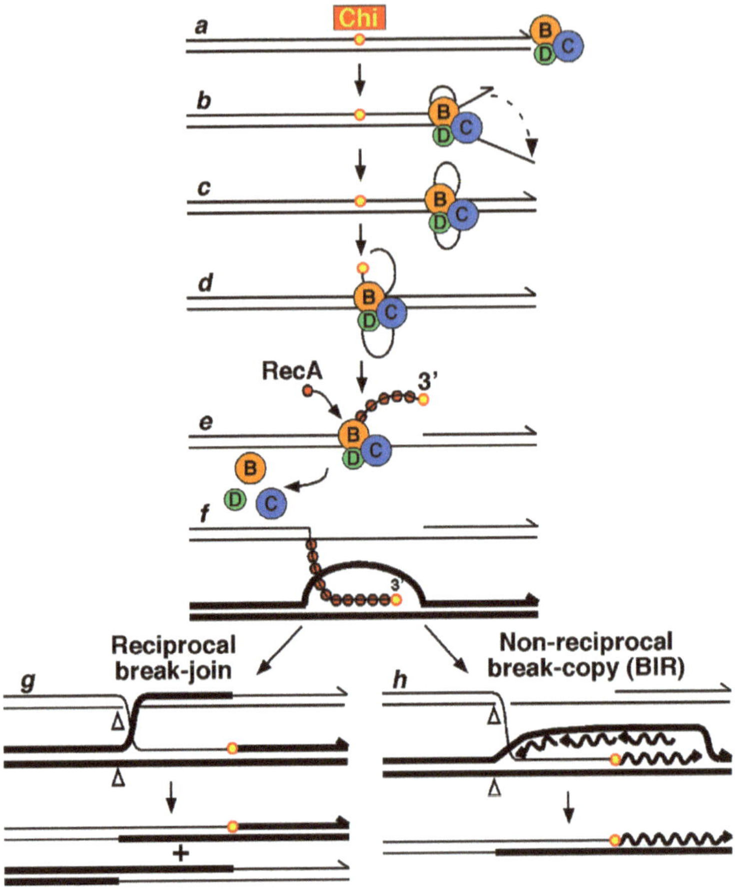

Fig. 1 Pathway of recombination promoted by Chi and RecBCD helicase-nuclease. (A) RecBCD binds the end of ds DNA (from linear DNA entering the cell or from a chromosomal break). (B) The RecD helicase moves along the 5′-ended strand faster than the RecB helicase moves along the 3′-ended strand, forming a loop-tail structure. (C) The tails anneal to form a twin-loop (aka "rabbit ear") structure. (D) When RecC encounters Chi on the 3′-ended strand emerging from the RecB helicase, the RecB nuclease cuts the 3′-ended strand about 5 nucleotides 3′ of the Chi octamer. (E) RecBCD loads RecA strand exchange protein onto the newly generated 3′-ended strand. At some point after acting at Chi, RecBCD dissociates into inactive subunits. (F) The RecA-ssDNA filament pairs with homologous dsDNA (thick line), such as the circular chromosome (only partially shown), forming a displacement (D-) loop. (G) The D-loop may be cut and form a Holliday junction, which can be resolved into reciprocal recombinants (cutting, swapping of ends, and ligation at open triangles). (H) Alternatively, the 3′ end in the D-loop may prime DNA synthesis and formation of a dual-strand replication fork, which can be resolved, as in (G), into one non-reciprocal recombinant (shown) and a parental-type (non-recombinant) DNA molecule (not shown). On linear DNA entering the cell, RecBCD acts

Table 1 Chi is frequent in foreign DNA ("non-self" to *E. coli*).

Species	Observed Chi sites/Mbp	Rank of Chi among ~65,000 octamers[a]	% (G+C)	"Expected" Chi sites/Mbp[b]	Observed/ "expected" Chi
E. coli phage P1 (temperate)	527	1	48	25	21.1
Klebsiella pneumoniae	345	63	57	50	6.9
Pseudomonas aeruginosa	332	352	67	77	4.3
Xanthomonas sp.	330	350	66	75	4.4
E. coli phage T3 (virulent)	314	9	50	31	10.1
Serratia sp.	257	148	55	44	5.8
Mycobacterium tuberculosis	241	493	66	75	3.2
Escherichia coli	**217**	**21**	**51**	**32**	**6.7**
Salmonella enterica	179	145	52	36	5.0
Vibrio harveyi	110	290	45	20	5.5

[a]Phages P1 and T3 have 55659 and 24755 unique octamers, respectively. Bacteria listed have 65,000 ± <1000 unique octamers each.
[b]Based on the % (G + C) and assuming random association of nucleotides.
Data are from DNA sequences whose sources are in Table S11. See also Tables S2 and S3.

than in *E. coli* itself (Tables 1, S2, and S3; Fig. S1). [We express Chi's "density" as the number of Chi sites per million base-pairs (Mbp) of DNA.] For example, *E. coli* strain K12 substrain MG1655 has 1008 Chi sites in its 4.64 Mbp genome (217/Mbp). But there are >300 Chi sites/Mbp in *Klebsiella pneumoniae* (another enteric bacterium, like *E. coli*, in the order Enterobacteriales) and in the more distant *Pseudomonas aeruginosa* (order Pseudomonadales) and a *Xanthomonas* sp. (order Xanthomonadales).

as shown at both ends. Option (G) would generate a recombinant circle and a linear form, which could be lost or recombine again. Option (H) would generate two circles or a dimeric circle, which would be resolved by XerCD at the *dif* site into two circles (Smith, 1991). *From Amundsen, S. K., Taylor, A. F., Reddy, M., & Smith, G. R. (2007). Intersubunit signaling in RecBCD enzyme, a complex protein machine regulated by Chi hot spots.* Genes & Development, 21, *3296–3307; Smith, G. R., Schultz, D. W., Taylor, A. F., & Triman, K. (1981). Chi sites, RecBC enzyme, and generalized recombination.* Stadler Genetics Symposium, 13, *25–37.*

The view noted above would lead to the conclusion that *P. aeruginosa* is more like *E. coli* than is *E. coli* itself. Even *Mycobacterium tuberculosis* (order Actinomycetales), genetically and physiologically very distant from the enteric bacteria, has higher Chi density (241/Mbp) than *E. coli*. That Chi is more frequent in the DNA of many foreign species precludes Chi from being the basis for *E. coli* telling DNA as foreign (non-self) or not (self). We see no mechanism by which any feature frequent in both group A and group B, but more frequent in group A than in group B, can be used by group B to distinguish itself from group A.

Some *E. coli* phages lack Chi, as the view above would expect, but others have Chi, in some cases at two- or three-times higher density across their genome than in *E. coli* DNA (Tables 1, S2, and S3; Fig. S1). The view above likely derives, at least in part, from phage lambda, with which Chi was discovered, lacking Chi in its 48.5 kbp of DNA. But this had to be the case, because Chi was first observed as a mutation that increased the plaque-size of certain lambda deletion mutants lacking its Red recombination pathway and Gam inhibitor of RecBCD (Henderson & Weil, 1975). If wild-type lambda had possessed Chi, Chi would not have been discovered in this way. Chi stimulates the *E. coli* RecBCD pathway to recombine monomeric lambda DNA to form dimeric (or higher order concatemeric) DNA essential for packaging lambda DNA into viable phage particles (Lam, Stahl, McMilin, & Stahl, 1974; Stahl & Stahl, 1977). Without Chi, this recombination occurs, but at low level; consequently, lambda *red gam* mutant plaques are small. With Chi's stimulation of recombination and phage propagation, the plaques are large. But in wild-type lambda (*red*$^{+}$ *gam*$^{+}$), plaque size is indistinguishable with and without Chi (Sprague, Faulds, & Smith, 1978). Note that the name Chi comes from its activity as a "crossover hotspot instigator" (Lam et al., 1974), but this key criterion is often overlooked when discussing Chi sites, as we note in section 5 below. Study of Chi in lambda for 50 years has been essential to working out the complex activities of RecBCD and its control by Chi (Smith, 2012).

Remarkably, some phages closely related to lambda do have Chi (Bobay, Touchon, & Rocha, 2013) (Tables S2 and S3). For example, four phages in Table S2 with lambda-size genomes and lambda-like gene organization have Chi densities similar to or even greater than that of *E. coli*. Two of these four phages encode apparent RecBCD inhibitors, like lambda's Gam protein, and two do not. Most of these Chi sites are in the orientation of active Chi in lambda, implying they are active in these native phages (Chi is active as a recombination hotspot in lambda only when correctly oriented with respect to the direction of its DNA packaging; Fig. 1 and see below).

Other well-studied phages unrelated to lambda, including both temperate and virulent types, also have Chi. Especially notable is temperate phage P1 (94.8 kbp DNA) and three relatives (P7, D6, and putative phage NGUA22), which have Chi density two or three times higher than that of *E. coli*. These phages may well use Chi to their advantage (see section 6 on temperate phage P1 below), but after most infections they kill *E. coli* despite having Chi. Three unrelated virulent phages (T3, T5, and T7) have Chi density near to or greater than that of *E. coli*. All these phages, obviously, succeed in killing *E. coli*: thus, Chi does not act as a signal of "self" DNA and thereby protect the cells from death. Even the virulent phages T2 and T4 have multiple Chi sites at "higher-than-expected" frequency (that based on random nucleotide association; Table S2), yet they kill *E. coli* soon after infection. We see no discrimination by *E. coli* between phages with Chi and those without. We note in section 8 below that phage lambda DNA broken by restriction enzymes is degraded by RecBCD whether or not it contains Chi (Simmon & Lederberg, 1972) and that *E. coli* DNA broken by irradiation is degraded by RecBCD (Clark et al., 1966) even though it has abundant Chi sites.

In a broader survey, we found that among 815 *E. coli* phages of diverse types, 720 (88%) contain Chi and 53 (6.5%) have Chi densities greater than that of *E. coli* (Table S3). 648 (80%) have more Chi sites than expected on the basis of random nucleotide association and their fraction of G+C, and 59% have more than twice as many Chi sites as expected (Fig. S1). Overall, the observed mean number is 3.22 times greater than the expected number (median = 2.75). Thus, phages have enriched their genomes for Chi, just as *E. coli* and other bacteria have (Table 1). The frequent presence of Chi among diverse phages, all of which kill *E. coli* in most infections, shows that Chi does not protect *E. coli* from these foreign (non-self) DNAs.

We know of no wild-type phage, with or without Chi, that is destroyed by RecBCD. Certain mutant phages, such as T4 *gene 2* mutants (Oliver & Goldberg, 1977), are destroyed by RecBCD despite having Chi, but such mutants are likely very rare in the natural world. Lambda *red gam* mutants, with or without Chi, are destroyed by RecBCD but only in *recA* mutant hosts (see sections 7 and 8 below on RecBCD destroying DNA that cannot recombine). As noted above, on wild-type *E. coli* wild-type lambda, without Chi, grows the same as lambda with Chi (Sprague et al., 1978). Thus, in the natural world we see no evidence that RecBCD destroys intracellular DNA because it does not have Chi, or that DNA is protected because it does have Chi.

Many plasmids, which replicate independently of the chromosome, do not normally kill the cell when they enter by conjugation (Johnson & Grossman, 2015). Many, such as *E. coli*'s 99.2 kbp F (fertility) factor, lack Chi, yet they are not destroyed upon entry and they are beneficial to *E. coli* by providing a means for gene exchange and rapid evolution. But other conjugative plasmids have Chi at high density. For example, plasmid pMG252, which confers quinolone-resistance and was first isolated from *K. pneumoniae* (Martinez-Martinez, Pascual, & Jacoby, 1998), has 81 Chi sites in its 186 kbp of DNA, or 436/Mbp, twice the density of Chi in *E. coli*'s chromosome. Plasmid pJP4, isolated from *Alcaligenes eutrophus*, even more distantly related to *E. coli* and conferring herbicide-resistance (Don & Pemberton, 1981), has 410 Chi sites/Mbp. Both plasmids move between a wide range of bacteria and belong to different incompatibility groups. So, *E. coli* does not treat plasmids as foreign or not depending upon their Chi content.

Parallel to the wide distribution of Chi, >90% of bacterial species with sequenced genomes encode RecBCD or an analog (AddAB or AdnAB) (Cromie, 2009). A recent study using stringent exclusion criteria found just three species of 128—from a survey spanning 33 bacterial phyla—that appeared to lack candidates for RecBCD, AddAB, and AdnAB (Gurung & Blumenthal, 2020). In most of the cases tested, these enzymes are important for DNA repair or recombination or both. Below, we discuss the hypothesis that Chi or other short sequences prompt these enzymes to promote recombination or enable them to distinguish self vs. non-self DNA.

3. Chi is active in "foreign" species

If Chi allows *E. coli* to tell self from non-self DNA, one would expect Chi to be active only in *E. coli*, but this is not the case. We note that two

Table 2 Chi is active beyond *E. coli*, in all tested enteric bacteria.

Species	**Chi nicking**[a]	**Chi hotspot activity**[b]
Escherichia coli	+++	7.3
Salmonella enterica	+++	3.7[c]
Klebsiella pneumoniae	+	4.1
Serratia marcescens	+	4.1
Proteus mirabilis	++	2.7
Vibrio harveyi	+++	ND
Pseudomonas aeruginosa	−	1.0
Pseudomonas putida	−	1.0

[a]Production of ssDNA cut at Chi during unwinding by RecBCD in extracts of native bacteria or of *E. coli* with cloned *recBCD* from other species (McKittrick & Smith, 1989; Schultz & Smith, 1986). +++, strong; ++, moderate; +, weak; −, undetectable.
[b]Enhancement of recombination in an interval with Chi in standard lambda hotspot crosses (Stahl & Stahl, 1977) in *E. coli* *ΔrecBCD* cells with cloned *recBCD* genes from the indicated species (McKittrick & Smith, 1989). A value of 1 indicates no enhancement (no Chi activity).
[c]3.2 in *S. enterica* cells (Smith et al., 1986).

divergent species must use different mechanisms to distinguish their DNAs. Extracts of other enteric bacteria, including *K. pneumoniae*, and even more distantly related *Proteus mirabilis* and *Vibrio harveyi* nick DNA in a Chi-specific manner, as do *E. coli* extracts or purified RecBCD (Table 2) (Schultz & Smith, 1986) (Fig. 1; see section 7 below on views about phage destruction for further discussion of RecBCD's action on DNA with Chi). When the *recBCD* genes of these enteric bacteria are expressed in *E. coli*, they confer Chi hotspot activity in genetic recombination assays and Chi-dependent nicking activity in enzymatic assays (McKittrick & Smith, 1989). Chi is also active in lambda crosses conducted in *Salmonella enterica* cells (Smith, Roberts, & Schultz, 1986). Thus, species other than *E. coli* have high density Chi sites, and they activate them as recombination hotspots. Clearly, Chi activity is not restricted to *E. coli* or even to the order Enterobacteriales.

By contrast, strong evidence indicates that Chi is not active in *Pseudomonas* spp. Extracts of *P. aeruginosa* and *P. putida* do not detectably nick at Chi, unlike extracts of all enteric bacteria tested (McKittrick & Smith, 1989; Schultz & Smith, 1986). Chi is not active as a recombination hotspot in *E. coli* cells expressing *recBCD* genes from either *Pseudomonas* species, but it is active in cells expressing each of the other enteric RecBCD enzymes tested (McKittrick & Smith, 1989). RecBCD enzyme purified from *P. syringae* cuts DNA at a different, but closely related, sequence 5' GCTGGCGC 3' (Pavankumar, Sinha, & Ray, 2018). This sequence may be a recombination hotspot in *P. syringae*, but to our knowledge this has not been directly demonstrated. If it indeed is a recombination hotspot, its sequence similarity to that of *E. coli* Chi provides an excellent opportunity for exploring the evolution of hotspots.

4. Is Chi "over-represented" in *E. coli*'s genome?

It is often stated, as part of the views discussed above, that Chi occurs in the *E. coli* genome more frequently than expected. For example, "Since Chi sites occur every ~5 kb in the *E. coli* genome, which is about 14 times more frequent than expected by chance, these sites were suggested as markers of bacterial self, ... [and are] responsible for some of the self/non-self discrimination properties of the CRISPR adaptation process" (Levy et al., 2015). See Table S1 for more examples. The basis of this expectation is not stated, but presumably it assumes random association of nucleotides in the *E. coli* genome. On this basis, many bacteria have more Chi sites than expected (Table 1). The nucleotide sequence of compact bacterial genomes, which must encode functional proteins, is not expected, however, to be random.

We therefore analyzed the *E. coli* genome for features that might account for Chi's density in *E. coli*. For the first analyses, we used only the "top" strand, that with 5' AGCTTTTCATTCTG 3' at nucleotides 1–14 in NC_000913.3, the sequence of *E. coli* strain K12, substrain MG1655. In this way, we might see asymmetries or other relations that would be missed using both strands. (Analysis of the bottom strand or both strands together gives similar results.) Table S4 gives for each nucleotide its count and frequency, which is close to, but not equal to, 0.25 for each nucleotide. Taking the observed frequencies and assuming random association, we calculate 149.9 expected Chi sites/9.28 million nucleotides (counting both strands, or 149.9/4.64 Mbp, in the *E. coli* genome). This is 6.72 times less than observed (1008/4.64 Mbp), as we noted previously (Colbert, Taylor, & Smith, 1998). (The value of 14 times quoted above may not have counted the nucleotides of both strands.) Thus, Chi is present more frequently than random nucleotide association predicts, leaving open the question of what, if anything, accounts for Chi's high frequency in *E. coli* DNA.

4.1 Nucleotides are not randomly arranged in the *E. coli* genome

We tested the randomness of nucleotides in the *E. coli* genome in various ways and observed high non-randomness. If nucleotides were randomly associated, the frequency of each dinucleotide (TT, TA, AT, etc.) would be the product of the individual nucleotide frequencies (Table S4). These calculated values are far from the observed values (Fig. 2A); counting dinucleotides as non-overlapping (1 2, 3 4, 5 6, etc.) or as overlapping (1 2, 2 3, 3 4, etc.) gave indistinguishable results (Fig. S2). Thus, neighboring nucleotides are highly correlated; but nucleotides 10000 apart, for example, are randomly associated (not significantly correlated) (Fig. 2B). We expected this result from consideration of neighboring nucleotides being correlated due to non-randomness of codon and amino acid frequencies (see next sections), protein-coding sequences (open reading frames, or ORFs) being less than 10000 nucleotides long (Fig. 3), and ORFs occupying 87.5% of *E. coli*'s genome. Beyond ORF lengths, we expected no nucleotide correlation, as observed (Fig. 2B). We then explored ORF features that might account for the observed frequency of Chi in the *E. coli* genome.

4.2 Amino acids are not quite randomly associated in *E. coli* proteins

The frequency of amino acids encoded by *E. coli* ORFs ranges from 10.7% for Leu to 1.2% for Cys (Table S5). The observed frequency of each dipeptide (LeuAla, ArgTrp, etc.) is similar to, but not exactly, that expected from random association (Fig. 4). Pearson's correlation coefficient is 0.98, but the slope of the linear regression line (0.94 ± 0.01) is

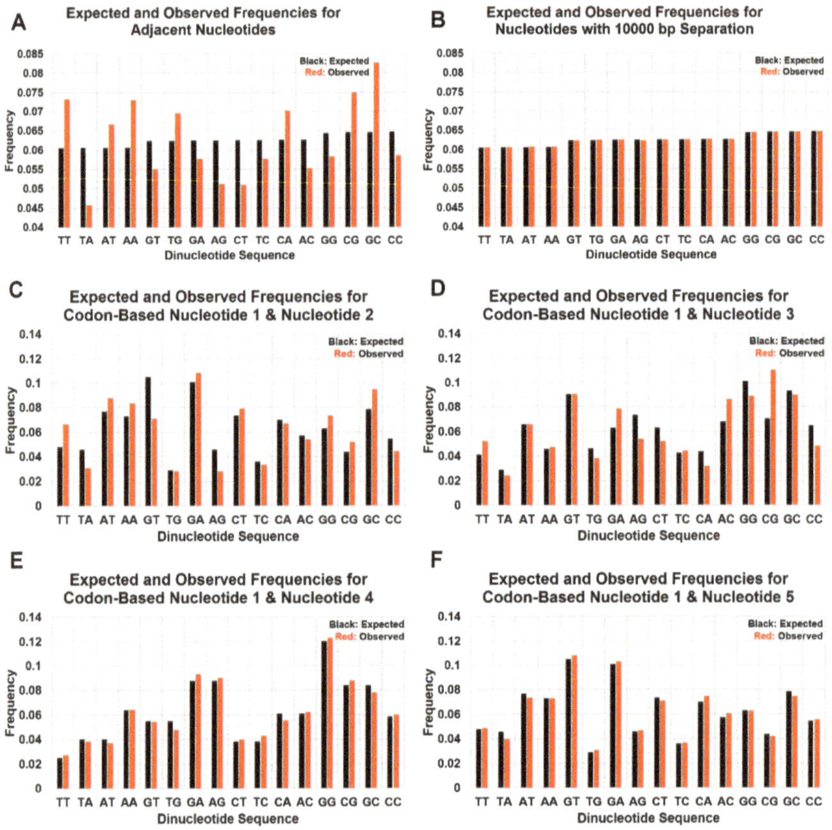

Fig. 2 Frequencies of close nucleotide pairs are not random, depend on the nucleotide positions within codons, and display a repeating pattern. Note differences in vertical scales. (A and B) The expected frequencies of each nucleotide pair, the product of the individual nucleotide frequencies in the *entire top strand* of *E. coli* (Table S4), are in black; in red are the observed frequencies for adjacent nucleotides (0 separation) (A) and for nucleotides separated by 10000 nucleotides (B). (C–F) The expected frequencies of each nucleotide pair, the product of the individual nucleotide frequencies at *each codon position* in all ORFs (Table 3), are shown for adjacent nucleotides (0 separation) (C) and for nucleotides separated by 1 (D), 2 (E), or 3 (F) nucleotides in black. Corresponding observed nucleotide pair frequencies are in red. Note that separations by 0, 1, or 2 nucleotides have different patterns, both expected and observed, but separation by 3 nucleotides (nucleotides in positions 1 and 5; F) resembles that of 0 nucleotide separation (nucleotides in positions 1 and 2; C).

highly significantly different from 1, the value that indicates random association ($P < 0.001$). Certain amino acid pairs occur noticeably more, and some less, frequently than expected from random association. For example, the observed and expected frequencies for ThrLeu are 0.723% and

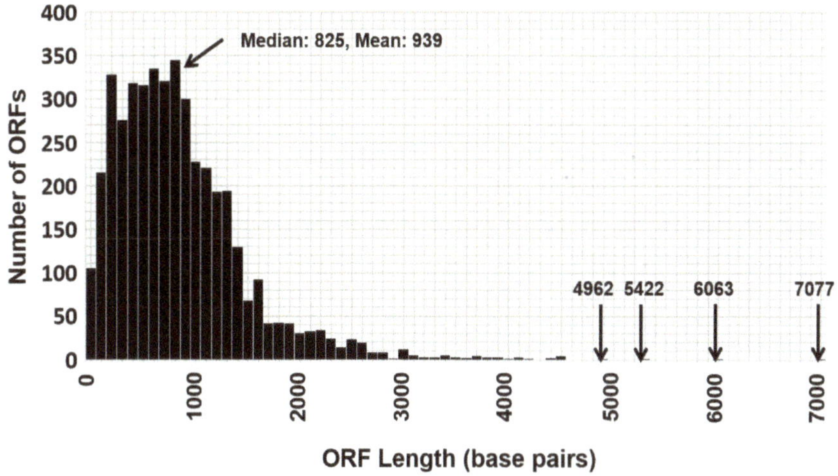

Fig. 3 The distribution of *E. coli* ORF lengths accounts for nucleotide associations becoming nearly random between about 1500 and 3500 bp. Shown are the number of ORFs (vertical axis) in each size range (bins of 100 bp; horizontal axis) and the four individual ORFs >4612 bp long. Figs. 2 and S4–S7 show that the non-randomness of adjacent nucleotides is greater than that of nucleotides separated by many nucleotides; the non-randomness becomes progressively less at 751, 1501, 3502, and 10,000 nucleotides of separation. This distance-dependence is accounted for by most ORFs being less than 1500 nucleotides long.

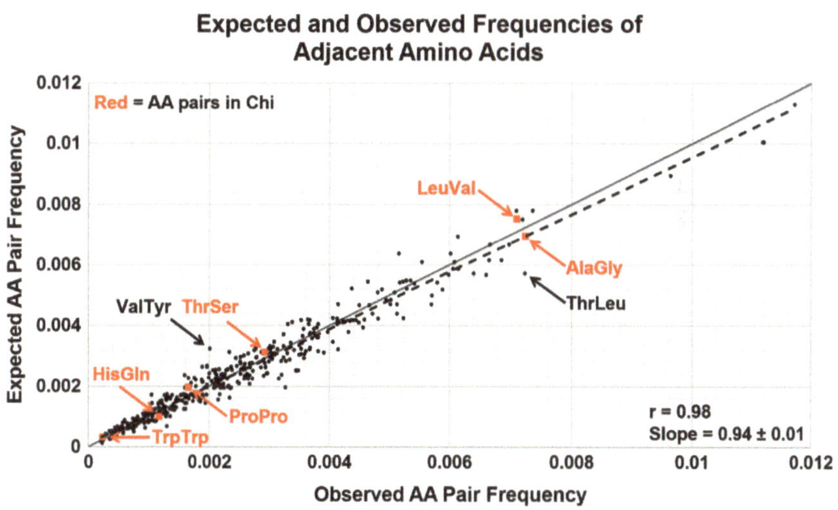

Fig. 4 *E. coli* amino acids are not quite randomly associated. The frequency of each pair of adjacent amino acids (AA) in all ORFs is plotted against its randomly expected frequency (product of individual amino acid frequencies in Table S5). Note that the slope of the linear regression line (dashed black) is significantly different from 1 (solid grey line, $P < 0.0001$ by two-sided unpaired t test), which would indicate randomness; Pearson correlation coefficient **r** is near unity, indicating high correlation. In red are dipeptides encoded by Chi or its complement. In black are two dipeptides whose observed frequencies are especially different from their expected frequencies ($P < 0.0001$ by two-tailed chi-square test). See also Fig. S3.

0.572%; those for ValTyr are 0.200% and 0.324% ($P < 0.001$ by two-tailed chi-square test). These results led us to examine codons for non-randomness.

4.3 Codon usage is even more non-random than amino acid usage in *E. coli*

Codon usage, like amino acid usage, varies highly. The most frequent codon, CTG for Leu, is used 45 times more frequently than the least frequent codon, AGG for Arg, even though both encode abundant amino acids (Tables 3 and S5). Even among codons for Leu, CTG is 14 times more

Table 3 Chi contains *E. coli*'s most frequent codon 5′ CTG 3′.

Codon	Number	Codon	Number	Codon	Number	Codon	Number
F[a] TTT	30335	*S* TCT	11527	*Y* TAT	22013	*C* TGT	7064
F TTC	22534	*S* TCC	11755	*Y* TAC	16670	*C* TGC	8871
L TTA	18957	*S* TCA	9780	* TAA	2875	* TGA	1416
L TTG	18659	*S* TCG	12209	* TAG	359	*W* **TGG**	20896
L CTT	15020	*P* CCT	9544	*H* CAT	17610	*R* CGT	28514
L CTC	15119	*P* CCC	7468	*H* CAC	13232	*R* CGC	30013
L CTA	5325	*P* CCA	11545	*Q* CAA	20957	*R* CGA	4885
L **CTG**	71938	*P* CCG	31705	*Q* CAG	39361	*R* CGG	7408
I ATT	41473	*T* ACT	12179	*N* AAT	24141	*S* AGT	11958
I ATC	34324	*T* ACC	31879	*N* AAC	29466	*S* AGC	21896
I ATA	5952	*T* ACA	9641	*K* AAA	45902	*R* AGA	2818
M ATG	37931	*T* ACG	19701	*K* AAG	14037	*R* AGG	1612
V GTT	24933	*A* **GCT**	20794	*D* GAT	43790	*G* **GGT**	33733
V GTC	20833	*A* GCC	34784	*D* GAC	26037	*G* GGC	40443
V GTA	14843	*A* GCA	27535	*E* GAA	53882	*G* GGA	10800
V **GTG**	35736	*A* GCG	45941	*E* GAG	24261	*G* GGG	15091
Total codons: 1363910 [4650 ORFs]							

[a]Letters in italics indicate the encoded amino acid, using the one-letter abbreviations in Table S5.
Data are for all ORFs ≥14 codons long as annotated in *E. coli* strain K12 substrain MG1655 (GenBank accession number NC_000913.3). Bold-face codons are in Chi (5′GCTGGTGG 3′).

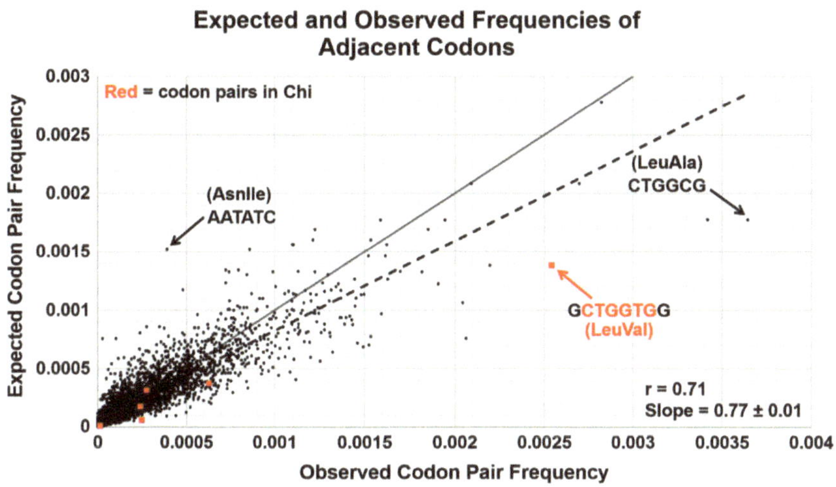

Fig. 5 *E. coli* codons are not randomly associated. The frequency of each pair of adjacent codons in all ORFs is plotted against its expected frequency (product of individual codon frequencies in Table 3). Note that the slope of the linear regression line (dashed black) is significantly different from 1 (solid grey line, $P < 0.0001$ by two-sided unpaired t test), which would indicate randomness; Pearson correlation coefficient **r** (0.71) also indicates significant but incomplete correlation. In red are dicodons contained in Chi or its complement. In black are two dicodons whose observed frequencies are especially different from their expected frequencies ($P < 0.0001$ by unpaired two-tailed chi-square test). See also Fig. S3.

frequent that CTA. (Note that CTG is part of Chi.) Codons are also non-randomly associated: the observed frequency of each codon pair is markedly different from their random association (Fig. 5), as noted by others (Boycheva, Chkodrov, & Ivanov, 2003; Gutman & Hatfield, 1989). (For calculations involving codons, we used all ORFs on both strands with $\geq$14 codons. ORFs on opposite strands only rarely overlap, providing non-redundant data.) Pearson's correlation coefficient is 0.71, and the slope of the linear regression line (0.77 ± 0.01) indicates high non-randomness. These not-surprising results led us to seek some feature of ORFs and codon usage that might account for the observed frequency of Chi.

We analyzed amino acids separated by one amino acid and found that these pairs are more nearly randomly associated than adjacent amino acids (0 separation) (Figs. S3A and S3C); pairs separated by two amino acids are even more nearly random. Similar relations occur for adjacent codons and those separated by one or two codons (Fig. S3B and S3D). These results show, as expected, progressively less correlation as amino acids or codons are farther apart and reinforce the non-randomness of the *E. coli* genome.

4.4 Preferential codon usage can account for Chi's high frequency in *E. coli*

We next asked whether the frequency of codon usage could account for the non-randomness of dinucleotides noted above. Although adjacent codons are not randomly associated (Fig. 5), we assumed randomness as a means to obtain "expected" values for comparison with the observed values in the following considerations. [ORFs occupy 42.5% of the top strand and 45.0% of the bottom strand; 99% (997/1008) of Chi sites are in ORFs. Thus, these calculations consider the genome portion containing nearly all Chi sites.] First, we calculated the frequency of dinucleotides expected from random association of codons. For this, we determined the frequency of each nucleotide at each position in codons (Table 4). We then calculated the frequency of each dinucleotide (TT, TA, AT etc.) at positions 1&2, 1&3, 1&4, ..., 2&3, 2&4, 2&5, ..., and 3&4, 3&5, 3&6, etc. expected from random nucleotide association and compared these with the observed frequencies (Fig. 2C–F). (We use "&" between nucleotides, e.g. 1&2 in Fig. 2C–F, when their positions in codons are considered and a space, e.g. 1 2 in Fig. 2A and B, when codon position is not considered.)

First, consider dinucleotides composed of nucleotides at the first and second positions (1&2) in each codon. The observed frequencies of TT, TA, AT etc. are similar, but not identical, to the expected frequencies (Fig. 2C). Thus, codon usage accounts for much of the non-randomness

Table 4 Chi (5′ GCTGGTGG 3′) contains *E. coli*'s "consensus" codon 5′ GTG 3′.

Nucleotide	Codon position 1	Codon position 2	Codon position 3
Thymine (T)	0.1583	**0.3035**	0.2600
Adenine (A)	0.2529	0.2893	0.1811
Guanine (G)	**0.3471**	0.1814	**0.2910**
Cytosine (C)	0.2417	0.2258	0.2679

Data (frequency of each nucleotide at each codon position) are from Table 3. The highest frequency at each position is in bold font. See also Colbert et al. (1998).

of nucleotides in the *E. coli* genome (compare Fig. 2A and C). Note that the range of observed frequencies considering ORFs and codon usage (from 0.0280 for TG to 0.108 for GA; Fig. 2C) is greater than the range observed (from 0.0457 for TA to 0.0828 for GC) for all dinucleotides without ORF or codon consideration; Fig. 2A). This codon-based analysis reveals an even stronger non-random feature of *E. coli*'s genome than the simple nucleotide-based analysis.

We similarly considered dinucleotides composed of the first and third positions (1&3) in each codon. Again, the observed frequencies in most cases are similar to the expected (Fig. 2D), but the pattern differs from that of positions 1&2 (Fig. 2C). Dinucleotides at positions 1&4 (i.e., the first position of adjacent codons; Fig. 2E) have a frequency distribution that differs even more than those of 1&2 or 1&3. The frequency distribution of 1&5 (Fig. 2F), however, is similar to that of 1&2, as expected if codon, and thus nucleotide, association becomes more nearly random with increasing distance. We quantified these comparisons (Fig. S4) and found, as expected, that the correlations between observed and expected frequencies are higher for identities (e.g., observed 1&2 compared with expected 1&2) (Fig. S4A) than for non-identities (e.g., 1&2 compared with 1&3) (Fig. S4B) except for 1&2 compared with 1&5 (Fig. S4C). These results confirm that codon usage accounts for much of *E. coli*'s non-random genome structure.

Going back to examination of nucleotide pairs on the top strand without consideration of ORFs or codons (Fig. 2A and B), we noted that the non-random pattern noted above persisted for quite some distance between nucleotides in the pair but became less noticeable in the range of several hundred bp (Figs. S5 and S6). There was, however, a 3 bp periodicity in the observed patterns, as expected from a major role for codon usage in determining the non-randomness discussed here and as previously noted in small *E. coli* phages (Shepherd, 1981). First, consider pairs of nucleotides separated by 2 or 5 nucleotides (e.g., nucleotides 1 4 or 1 7), which would be in the same codon positions (i.e., first, second or third position) if they are in the same ORF (there are no known introns in *E. coli* genes). The most frequent dinucleotides separated by 2 nucleotides are also the most frequent dinucleotides separated by 5 nucleotides; there is a similar pattern for the least frequent dinucleotides, although in each comparison the magnitudes differ (Fig. S6A and D). Thus, the observed patterns of separation by 2 or 5 nucleotides are related. Furthermore, separations by 3 or 6 nucleotides are similar (Fig. S6B and E), as are separations by 4 or 7 nucleotides

(Fig. S6C and F). At 749 – 754 nucleotide separations, the dinucleotide frequency distributions are still non-random but not as highly non-random as with less separation (e.g., 0, 2, 3, or 4 nucleotides; Fig. 2A and S6). The pattern with separation of 749 nucleotides (e.g., 1 751) is similar to that with separation of 752 nucleotides (e.g., 1 754), just as 1 4 is similar to 1 7, etc. (Fig. S6A and D). These 3-nucleotide periodicities are expected from the influence of codon usage. The pattern becomes more nearly like that of random nucleotide association (Fig. 2A) as the separation increases to 1501 and 3502 nucleotides (Fig. S5 and S7). With 10000 nucleotides of separation (Fig. 2B), the pattern is indistinguishable from that of random association, as noted before. These outcomes are expected from the distribution of *E. coli* ORF lengths (Fig. 3).

These analyses show that the non-randomness of *E. coli*'s genome stems from codon usage. They support the view that the often-stated "unexpectedly" or "disproportionately" high frequency or "overrepresentation" of Chi is due to its codon usage, as noted by others and one of us (Biaudet, El Karoui, & Gruss, 1998; Colbert et al., 1998; Uno, Nakayama, & Tomita, 2006), rather than recombination-based selection enriching this octamer. We redid the calculations by Colbert et al. (1998) for the expected and observed frequency of Chi in each reading frame (Table 5) using a slightly different genome sequence and set of ORFs for

Table 5 Chi's frequency is correlated with its in-frame dicodon frequencies.

Dicodon	Expected number[a]	Observed number	Proportion of observed (%)	Proportion of Chi sites in this frame (%)
GCTGGT	519	542	9.44	5.82
CTGGTG	1901	3445	60.03	68.21
TGGTGG	321	334	5.82	5.42
Subtotal	2741	4321		
CCACCA	98	126	2.20	1.54
CACCAG	385	392	6.83	4.81
ACCAGC	516	900	15.68	14.24
Subtotal	999	1418		
Total	3740	5739	100	100

[a]Based on random codon association, calculated using values from Table 3. See also Colbert et al. (1998).

codon usage. The results confirm that Chi occurs most often with CTG in the reading frame, as expected from CTG being the most frequent codon for the most frequent amino acid (Leu) (Tables 3 and S5). The expected frequencies of Chi or its complement (5' CCACCAGC 3') in each reading frame are similar to the observed frequencies. Thus, the observed frequency of Chi can be nearly fully accounted for by its sequence, *E. coli*'s codon usage, and ORFs occupying most of the genome.

Chi's non-random orientation on the chromosome can also be accounted for by *E. coli*'s non-random orientation of ORFs (Bell, Chow, Ho, & Forsdyke, 1998; Colbert et al., 1998). Chi is oriented on the *E. coli* genome in a manner consistent with Chi being active in repair of DNA broken during replication. When a replication fork is broken on either the leading or lagging strand, RecBCD can enter the two ds DNA ends and promote restoration of the fork (Kuzminov, Schabtach, & Stahl, 1994). Entry onto the end heading RecBCD toward the origin of replication, but not onto the other end, would lead to the entered end being rejoined to the intact DNA homologous to it and allow replication to restart at the resultant new fork. Chi oriented with the entry end to the right of 5' GCTGGTGG 3' would activate RecBCD when it encounters Chi (Taylor, Schultz, Ponticelli, & Smith, 1985). 75.5% of Chi sites are oriented in this way (replication proceeds in both directions from the unique origin). Genes, especially highly expressed, essential ones, are preferentially oriented such that they are transcribed, and accordingly translated, in the same direction as replication, presumably to avoid head-on collisions of transcription and replication (Brewer, 1988; Rocha, 2004; Schroeder, Sankar, Wang, & Simmons, 2020). These biases thus orient Chi in the direction favoring repair of broken replication forks. It follows that the biased orientation of Chi is no mystery—it stems from transcription, and consequently translation, being biased in the direction of replication plus non-random codon usage. Note that this role for Chi and RecBCD is repair of self DNA, not destruction of non-self DNA.

Buton & Bobay (2021) attempted to predict the Chi-equivalent sequence in other Proteobacteria, based on assumptions of similarity to Chi's sequence, high frequency, and chromosomal orientation-dependence. They failed to detect the "Chi" sequence of *H. influenzae* noted in section 5 below, suggesting that these computational methods are inadequate. Perhaps only experimental searches, such as those leading to discovery of *E. coli*'s Chi, will successfully identify recombination hotspots.

What led to Chi being 5' GCTGGTGG 3'? It is clearly highly represented in the genome—Chi and its complement are the 21st and 22nd most common octamers of 65,484 present on the two strands of *E. coli* DNA (Fig. 6). The "consensus codon" GTG, that with the most frequent nucleotide at each position (Table 4), and CTG, the most frequent codon (Table 3), are both in Chi. 5' CTG|GTG 3', the central part of Chi, contains the most frequent codons for Leu and Val, two of the four most-frequent amino acids encoded by *E. coli* (Tables 3 and S5) and is the fifth most abundant dicodon, among 3721 dicodons (Fig. 5). These features likely have something to do with 5' GCTGGTGG 3' being "chosen" as Chi. But these are features of translation, not of recombination. We suppose that an ancient ancestor of *E. coli* had a RecBCD-like DNA repair enzyme with no sequence preference. A *recBCD* mutant with weak preference for some DNA sequence arose and had a higher efficiency of repair. Additional mutations with increasing efficiency gained abundance in the population. One that used a highly frequent sequence, such as 5' GCTGGTGG 3', would have been the most efficient and abundant. Perhaps *E. coli* is still struggling to recognize the most frequent octamer 5' CGCTGGCG 3' in its genome.

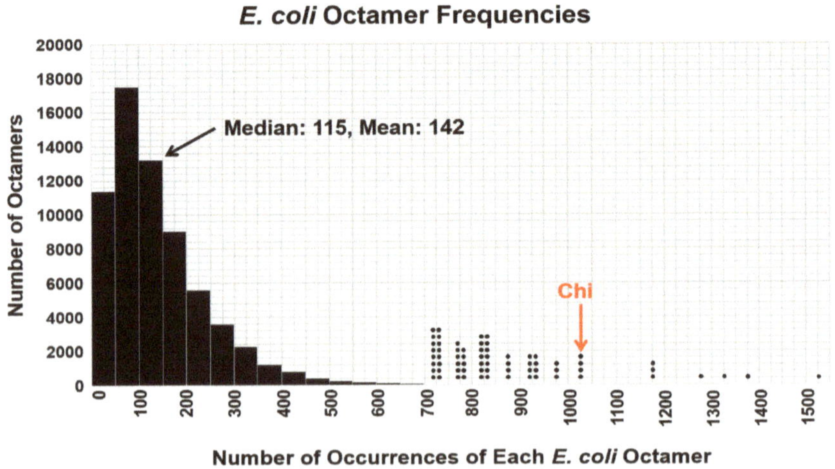

Fig. 6 Chi is one of the most frequent *E. coli* octamers. The number of occurrences in the *E. coli* genome (both strands) of each octamer-complement pair, in bins of 50 (horizontal axis), is plotted against the number of octamer-complement pairs in that frequency range (vertical axis). There are 65484 octamers in the genome. Those occurring >700 times each are shown by individual bullets (●), one for each octamer-complement pair. Chi and its complement (red arrow) occur 1008 times.

It is also possible that the opposite is true: perhaps RecBCD first developed affinity for Chi, and then *E. coli* evolved to increase the frequency of Chi in its genome, for instance by enriching for silent mutations that make Leu codons into CTG. This possibility, however, requires that codon usage (protein synthesis), which occurs $\sim 10^9$ times per generation (Milo, 2013), evolved to accommodate a rarely used feature (recombination or DNA break repair), which occurs about once per generation, and thus seems less likely.

5. "Chi" sequence of *S. aureus* also comports with its codon usage

The concepts developed here can also account for the sequence 5' GAAGCGG 3' presumably affecting the activity of the *Staphylococcus aureus* AddAB enzyme, its RecBCD analog (Halpern et al., 2007). This sequence specificity was identified from studies of rolling circle replication. Some phages, including lambda, and some plasmids change from circular (theta, θ) replication to rolling circle (sigma, σ) replication during their infection cycles. The end of the rolling circle can be degraded by RecBCD or AddAB nuclease activity. For example, *S. aureus* plasmid pC194 replicating as a rolling circle in *E. coli* cells forms high molecular weight (HMW) DNA only if the plasmid contains Chi, providing a convenient screen for DNA containing a special sequence (Dabert, Ehrlich, & Gruss, 1992). In *S. aureus*, this plasmid makes HMW DNA if it contains 5' GAAGCGG 3' (Halpern et al., 2007). Thus, this sequence is often considered to be *S. aureus*'s "Chi" (but see next paragraph). This heptamer and its complement occur 335 times in the *S. aureus* 2.82 Mbp genome and are the 33rd and 34th most frequent heptamers. (By contrast, Chi 5' GCTGGTGG 3' ranks 21,806 among *S. aureus*'s 65,327 octamers.) 89.7% of these "Chi" sequences are co-oriented with replication and 92.8% occur in ORFs. *S. aureus*'s codon usage can account for the frequency of *S. aureus*'s "Chi," just as for *E. coli*'s Chi (Tables S6 and S7). Thus, this sequence shares many properties with Chi and may be a recombination hotspot, but to our knowledge that feature and direct interaction with AddAB have not been demonstrated.

Other sequences that stimulate HMW DNA formation of pC194 have been found for other species and have been called "Chi." These species and their "Chi" sequences include *B. subtilis* (5' AGCGG 3', the 3' end of the *S. aureus* "Chi" sequence), *Lactococcus lactis* (5' GCGCGTG 3'), and *Haemophilus influenzae* [5 'G(G/C)TGGAGG 3' and 5' GNTGGTGG 3', both similar to *E. coli* Chi]. Each of these has been called a hotspot of recombination (Biswas et al., 1995; Chedin, Noirot, Biaudet, & Ehrlich, 1998; El Karoui, Ehrlich, & Gruss, 1998; Sourice, Biaudet, El Karoui, Ehrlich, & Gruss, 1998), but we are unaware of any recombination assays to demonstrate this property. Purified *B. subtilis* AddAB does cut DNA at its "Chi" site, showing, as expected, a direct interaction with this recombination-promoting enzyme (Chedin, Ehrlich, & Kowalczykowski, 2000). To our knowledge, a direct interaction between these "Chi" sequences and *L. lactis* AddAB or *H. influenzae* RecBCD has not been reported. Furthermore, *H. influenzae* "Chi" activity in HMW formation is not orientation-dependent, unlike the other Chi and "Chi" sequences discussed above. The formation of HMW DNA is often interpreted as "Chi" blocking the nuclease activity of the AddAB or RecBCD enzyme (Chedin et al., 1998; Dabert et al., 1992; Halpern et al., 2007; Sourice et al., 1998). But this simple view does not explain why formation of HMW DNA requires RecA protein (see sections 7 and 8 below about RecBCD destroying DNA).

6. Temperate phage P1 contains 50 Chi sites, likely uses them to its advantage, and appears to select for Chi as a recombination hotspot

We noted above that phage P1 contains Chi at more than twice the density of *E. coli* (Table 1). P1 has 50 Chi sites across its 94.8 kbp genome (Fig. 7). These Chi sites are abundant on both strands (31 are on the "top" strand; Table S8), but their positions are clearly non-random. Chi is much denser toward the "ends" of the genome than in the middle of the genome. We understand this non-random distribution as follows. During its lytic phase, P1 replicates to make concatemers about four units long (Sternberg & Hoess, 1983). It initiates packaging by cutting the DNA at its unique *pac* site and proceeding in one direction (to the right as shown in Fig. 7). When it reaches the intact *pac* site one genome length away, it proceeds ~5 to 10 kbp farther and cuts again (Iida & Arber, 1979; Ikeda & Tomizawa, 1968; Yun & Vapnek, 1977). This linear DNA, with terminal repeats, is packaged into the phage particle (virion). The next phage DNA is about the same length as the first and has ~5 to 10 kbp terminal

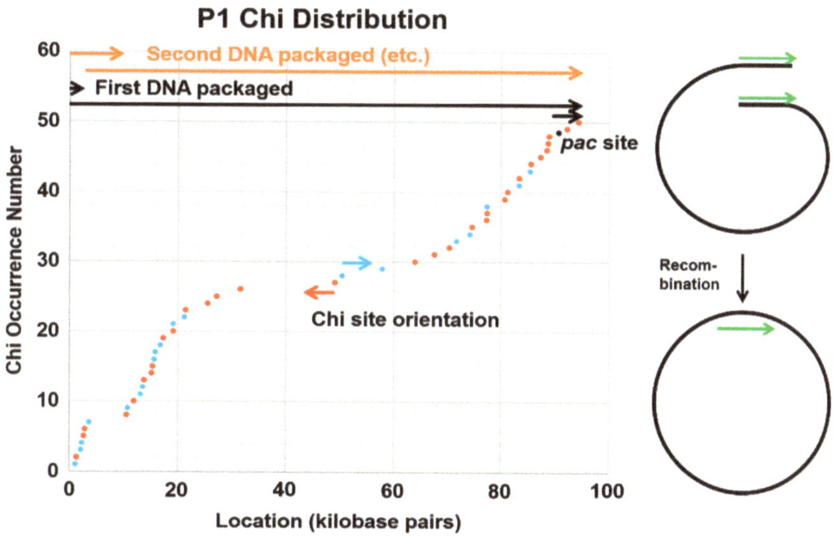

Fig. 7 High Chi-density in *E. coli* phage P1 can help RecBCD circularize P1's injected linear DNA. Chi sites are placed according to their numerical order from the left end on the unique P1 genome sequence. Chi sites at red dots activate RecBCD unwinding DNA from a ds end to its right; those at blue dots activate RecBCD from the left. Note the higher Chi density toward each end. See section 6 for description of sequential DNA packaging beginning at the unique *pac* site (black dot). Diagram on the right illustrates recombination between the terminal direct repeats (~5–10 kb; green arrows) to form circular DNA essential both for lysogenic propagation as a plasmid and for lytic growth to produce concatemers for packaging into virions. We propose that P1 and P7 use Chi sites to the right of the *pac* site, as shown here, and D6 and NGUA22, with inverted *pac* sites, use Chi sites to the left, to promote DNA circularization.

repeats of the genome sequence adjacent to the first terminal repeats. Packaging proceeds similarly for about four units. Thus, the far end of the fourth packaged DNA is ~20 to 40 kbp from the initial *pac* site. Most of the phage DNA ends are thus in genomic regions with dense Chi sites (Fig. 7).

Upon injection into the next (recipient) cell, the terminal repeats must recombine to form circular DNA (Sternberg & Hoess, 1983). This DNA can either undergo rolling circle replication in a lytic infection or be replicated as a circular plasmid in a lysogenic infection, the two lifestyles of P1. Circularization requires the host RecBCD pathway: P1 replication and phage yields are reduced ~20-fold and the phage do not form visible plaques in *recA* or *recBC* null mutants or in *recC* mutants altered in the RecBCD tunnel in which Chi is recognized (Amundsen, Sharp, & Smith, 2016; Zabrovitz, Legev, & Cohen, 1977). Thus, Chi-stimulated recombination is essential for P1 growth, likely at the circularization step. The abundant Chi sites, distributed as they are, ensure that RecBCD entering from either end of the injected linear DNA encounters an actively oriented Chi site. [Chi is active only when RecBCD approaches 5' GCTGGTGG 3' from the right as written here (Faulds, Dower, Stahl, & Stahl, 1979; Taylor et al., 1985).] It is not surprising, then, that Chi is the most frequent octamer among the 44458 (top strand) or 55659 (both strands) octamers present in P1.

Several observations lead us to propose that Chi's high frequency and distribution in the genome of P1 and its related phages reflect Chi being selected during evolution as a hotspot to promote recombination (circularization of the linear DNA after injection; Fig. 7). This differs from the situation in *E. coli*, where we infer that selection for codon usage results in Chi's high frequency. The following observations support this proposal.

(1) P1 requires RecA-RecBCD-promoted recombination for growth (Zabrovitz et al., 1977), but *E. coli* does not.

(2) Chi is located preferentially near the ends of P1's genome (Fig. 7), where recombination is required, but Chi is nearly uniformly distributed across *E. coli*'s genome although, like its ORFs, Chi is preferentially located on the leading strand of replication.

(3) Chi is the most frequent octamer in P1 but is only the 21st-most frequent in *E. coli* (Table 1), suggesting a stronger or more direct selection for Chi in P1 than in *E. coli*.

(4) Chi is much denser in non-coding DNA (non-ORFs) in P1 than in *E. coli* ($P = 0.0041$; Table S8). In P1 Chi occurs at nearly the same density in ORFs and non-ORF sequences ($P = 0.44$), but it is 13 times more frequent in *E. coli* ORFs than non-ORFs ($P < 0.0001$; Table S8).

(5) When Chi is in ORFs, the coding strand contains 5' GCTGGTGG 3' less often in P1 (63%) than in *E. coli* (79%) ($P = 0.015$; Table S8), indicating a closer connection between Chi and translation in *E. coli* than in P1.

(6) CTG, part of this sequence, is the most frequent codon in *E. coli* but is only the fourth most frequent codon in P1 (Tables 3 and S9). The frequency of CTG, encoding Leu, is 3.2 times the mean of all codon frequencies in *E. coli* but only 1.7 times the mean in P1 ($P < 0.0001$ by two-tailed chi-square test), even though Leu is nearly as frequent in P1 (8.7%) as in *E. coli* (10.7%); Leu is the most frequent amino acid in *E. coli* and the second most frequent in P1 (Tables 3, S5, and S9). CTG is 50% of all Leu codons in *E. coli* but only 32% in P1 ($P < 0.0001$ by two-tailed chi-square test). Other codons are used at more nearly equal frequencies in the two genomes (Fig. S8).

(7) In *E. coli*, 68% of all Chi sites in ORFs are in the reading frame 5' CTG|GTG 3', significantly more than in P1 (41%, $P < 0.0003$ by two-tailed Fisher's exact test; Tables 5 and S10), another indication that Chi is more closely associated with coding in *E. coli* than in P1.

Phages P7 (a close P1 relative) and D6 and putative phage NGUA22 [each distantly related to P1 and to each other (Gilcrease & Casjens, 2018); Table S8] also share these features with P1—Chi is the most frequent octamer in their genomes, it is more abundant near their genome ends, it is nearly randomly oriented on their genomes, they presumably require RecA and RecBCD to circularize their genomes for growth, they use codon CTG about half as frequently as *E. coli* does, and their Chi sites are in the reading frame 5' CTG|GTG 3' less often than in *E. coli* (Table S10). In other words, P1 and its relatives put Chi toward the genome ends (for recombination), and *E. coli* puts it in ORFs (for protein coding). This proposal and evidence for it are, to our knowledge, the first pointing toward a direct selection during evolution for Chi as a recombination hotspot.

7. A myth: "Chi converts RecBCD from phage destruction to DNA repair"

This statement (Cheng et al., 2020) and similar ones so often written (Table S1) would be compatible with the prevailing views (dogmas) discussed in the Introduction if it were true. But the mere growth, in *recBCD*$^+$ *E. coli*, of phage without Chi (Tables S2 and S3) shows that phages without Chi are not destroyed by RecBCD. Among known phages, this view could not hold, for to our knowledge only *recBCD*$^+$ bacteria have been used as hosts to search for *E. coli* phages (G. Hatfull, personal communication). If RecBCD destroyed a phage, it would not have been observed and described. (It might be interesting, however, to search for phages that do grow only in *recBCD* mutants. Their behavior would almost certainly lead to new insights into phage and RecBCD physiology.)

The claim above likely stems from purified RecBCD's activity at low ATP concentrations (less than that of Mg^{2+} ions). ATP chelates Mg^{2+}, and unchelated ("free") Mg^{2+} is required for high RecBCD nuclease activity (Rosamond, Telander, & Linn, 1979). With excess ATP, the free Mg^{2+} concentration is low and the nuclease activity is low; with excess Mg^{2+}, the nuclease activity is high (Murphy, 2000; Wright, Buttin, & Hurwitz, 1971). Thus, with excess ATP, RecBCD unwinds DNA from an end and simply nicks DNA at Chi, about five nucleotides to the 3' side of 5' GCTGGTGG 3' (Taylor et al., 1985; Taylor & Smith, 1995) (Fig. 1). With excess Mg^{2+}, RecBCD frequently nicks DNA before encountering Chi [forming single-stranded (ss) DNA a few kb long], nicks at and near Chi (within and to both sides of 5' GCTGGTGG 3'), nicks the complementary strand within and to both sides of 5' CCACCAGC 3', and nicks that strand during continued unwinding (Anderson & Kowalczykowski, 1997a; Dixon & Kowalczykowski, 1993; Taylor & Smith, 1995). Deducing which reaction occurs in *E. coli* cells is limited by lack of knowledge of the effective (free) concentration of Mg^{2+} in cells. Although in cells total Mg^{2+} (~100 mM) is greater than total ATP (~3 mM), most of the Mg^{2+} is likely bound to ribosomal RNA (see Smith, 2012; Taylor & Smith, 1995). The effective (free) concentration of Mg^{2+} in *E. coli* cells is uncertain from direct biochemical measurements. Thus, it is important to turn to other information to deduce the reaction of RecBCD before, at, and after Chi in living cells.

Genetic evidence, coupled with the biochemical activities of purified RecBCD, strongly indicates that *in living E. coli cells RecBCD nicks DNA at Chi*, as purified RecBCD does with excess ATP. The activity of Chi is strongly influenced by nucleotides 4, 5, and 6 to the 3' side of the Chi octamer 5' GCTGGTGG 3' (Amundsen et al., 2016; Taylor, Amundsen, & Smith, 2016). This is true for genetic experiments in cells; it is also true for biochemical experiments with purified RecBCD when ATP is in excess but *not* when Mg^{2+} is in excess. Additional evidence also indicates simple nicking at Chi in cells (Smith, 2012). For example, Chi acts in *trans* (i.e., between DNA molecules) in cells (Köppen, Krobitsch, Thoms, & Wackernagel, 1995; Myers, Kuzminov, & Stahl, 1995) and with purified components with excess ATP but *not* with excess Mg^{2+} (Taylor & Smith, 1992, 1999). Certain *recC* mutants lack RecBCD exonuclease activity but nevertheless retain Chi hotspot activity in cells, and their extracts nick at Chi (Amundsen & Smith, 2007). Furthermore, in lambda single-burst analyses, RecBCD pathway recombination is reciprocal (Sarthy & Meselson, 1976): ++ crossed with *ab* produces both +*b* and *a*+ at nearly equal frequencies in one infected cell. When Chi is between *a* and *b* and stimulates their recombination, degradation up to Chi, as envisaged by the widely-held view, would destroy one of the flanking alleles and not allow formation of reciprocal recombinants. Others report Chi-promoted recombination to be either reciprocal (Kobayashi, Stahl, Fairfield, & Stahl, 1984; Stahl, Lieb, & Stahl, 1984; Stahl, Stahl, Young, & Kobayashi, 1982), non-reciprocal (Lam et al., 1974; Stahl, Stahl, Malone, & Crasemann, 1980), or both (Ennis, Amundsen, & Smith, 1987; Stahl, Shurvinton, Thomason, Hill, & Stahl, 1995; Stahl, Thomason, Siddiqi, & Stahl, 1990). Both reciprocal and non-reciprocal recombination promoted by RecBCD and Chi can be accounted for by the nick-at-Chi model (Fig. 1). We are unaware of corresponding, consistent arguments for degradation by RecBCD up to Chi *in living cells*. Thus, the available evidence strongly supports simple nicking at Chi.

8. An observation: RecBCD destroys DNA in *E. coli* only if it cannot recombine

A widely-held view (dogma) holds that RecBCD destroys DNA that does not contain Chi. The growth of phages without Chi or RecBCD inhibitors, discussed above, indicates this view is incorrect. Additional evidence comes from Chi's stimulation of HMW plasmid DNA in *E. coli*, noted above (Dabert et al., 1992). Formation of HMW DNA requires RecA DNA strand-exchange protein, an essential recombination factor that acts after RecBCD and Chi (Fig. 1). RecA binds ss DNA and promotes formation of joint molecules with homologous ds DNA; joint molecules are resolved into recombinants, the products of DNA break repair. Purified RecBCD loads RecA onto the ss DNA produced by RecBCD after it cuts at Chi (Anderson & Kowalczykowski, 1997b). When RecBCD encounters a properly oriented Chi on the linear "tail" of a rolling circle, the RecA-ssDNA complex can form a joint molecule with another plasmid in the cell, which can lead to a dumbbell-shaped molecule without ends; i.e., the ds DNA end for RecBCD's entry is removed by recombination (Smith, 2012). The plasmid is thus protected from cellular exonucleases, including the dozens of other RecBCD molecules in the cell. Continued replication produces HMW DNA. Without RecA, RecBCD can still nick DNA at Chi on the rolling circle, but without RecA's formation of joint molecules and recombinants, other RecBCD molecules or nucleases eventually destroy the DNA. [On ss DNA, RecBCD nuclease is highly active (Wright et al., 1971) but does not respond to Chi (Ponticelli, Schultz, Taylor, & Smith, 1985).] Thus, Chi's stimulation of HMW DNA formation cannot be taken as evidence that Chi switches RecBCD from a destructive nuclease to a beneficial repair enzyme. Coupled with RecA, the statement is true, but it is RecA-dependent recombination, not just Chi recognition by RecBCD, that blocks DNA destruction.

A different assay of DNA in *E. coli* leads to the same conclusion. Survival of plasmid DNA cut at a unique site is increased by a properly oriented Chi site and even more by two, tandem Chi sites (Kuzminov et al., 1994). As above, survival depends on RecA and ss DNA binding protein (SSB), which aids RecA in formation of joint DNA molecules. The surviving DNA contains branched DNA, likely arising from joint DNA molecule formation by RecA and SSB. If Chi inactivated RecBCD nuclease, this requirement for RecA would not be expected. As noted above, it is recombination, not just Chi, that prevents DNA destruction. In addition, plasmid concatemers are detected in recombination-proficient *recD* mutants, which lack RecBCD nuclease activity (Amundsen, Taylor, Chaudhury, & Smith, 1986), but only if RecA is present (Biek & Cohen, 1986), implying that other nucleases also can destroy DNA that does not recombine.

Growth of a phage is often assayed by plaque formation, which arises from a single phage particle infecting a cell. In this case, there is no homologous DNA with which the phage, or linearized form of it, can recombine to prevent eventual destruction. For example, RecBCD rapidly solubilizes about 40% of lambda DNA at a multiplicity of infection (MOI) of 1, a condition under which about 40% of the cells have no homolog with which lambda can recombine (Simmon & Lederberg, 1972). That is, a second (intact) DNA and presumably recombination are required for protection. Furthermore, after cleavage of lambda DNA by a cellular restriction enzyme, degradation by RecBCD is at least as rapid on DNA with Chi as on DNA without Chi, and *E. coli* DNA, replete with Chi sites, is degraded by RecBCD after irradiation in a *recA* mutant (Clark et al., 1966). Thus, clear evidence for Chi converting RecBCD from a destructive mode to a repair mode in cells is lacking.

9. Mechanisms employing RecBCD that do distinguish self vs. non-self DNA

If Chi does not enable bacteria to tell self from non-self DNA, are there mechanisms that do? Yes—several mechanisms have been described, including some involving RecBCD but not in the manner envisaged by the widely-held views discussed above. [See Rostol & Marraffini (2019) for additional mechanisms not involving RecBCD.] Perhaps the most fundamental basis for self vs. non-self determination is nucleotide sequence divergence among species. Divergent (heterologous) DNA undergoes RecA-promoted DNA strand exchange and thus RecBCD-promoted genetic recombination less frequently than nearly identical (homologous) DNA; this effect is independent of Chi (Rao, Chiu, Bazemore, Reddy, & Radding, 1995).

The first-described mechanism of self vs. non-self determination is restriction and modification by enzymes now universally used for DNA manipulation (Murray, 2002). In initial observations, phage T2 grown in *E. coli* strain X produced only a few phage that grew in another strain Y (i.e., it was restricted by strain Y), but the rare phage that had grown in strain Y grew in both strain X and strain Y (i.e., it was modified by strain Y); similar observations were made with other phages and bacterial strains (Luria, 1953). We now know that nearly all bacteria modify their own DNA and that of lucky phage escapees by methylation of the DNA at special sequences. Most restriction enzymes cut DNA that is not modified (not methylated, non-self DNA) at these sites but do not cut modified (methylated, self) DNA. As noted above, phage lambda cut by *Eco*KI restriction enzyme is further degraded by RecBCD, whether or not it contains Chi (Simmon & Lederberg, 1972).

A more complex mechanism, CRISPR-Cas, was described more recently (Hille et al., 2018). About half of bacterial genomes examined contain *C*lustered, *R*egularly *I*nterspersed, *P*alindromic *R*epeats (CRISPR). The repeats are ~30 bp long and can form a "snap-back" internal hairpin; they are separated by ~35 bp of unique sequence DNA (spacers). These "islands" usually contain <50, but occasionally >500, such alternating repeats and spacers. They remained mysterious for about 20 years after their discovery in *E. coli* (see Ishino, Krupovic, & Forterre, 2018; Ishino, Shinagawa, Makino, Amemura, & Nakata, 1987), when it was noted that the spacer DNA sequences often matched those of phages or plasmids (i.e., foreign or non-self DNA) (Bolotin, Quinquis, Sorokin, & Ehrlich, 2005; Mojica, Diez-Villasenor, Garcia-Martinez, & Soria, 2005; Pourcel, Salvignol, & Vergnaud, 2005). This observation gave rise to the idea that CRISPR islands are involved in phage defense, or self vs. non-self recognition. Further experiments in *Streptococcus thermophilus* showed directly that infecting phage DNA could be integrated as new spacers in the CRISPR locus and that plasmid or phage DNA could be cut at a predetermined site upon transformation or infection (Barrangou et al., 2007; Garneau et al., 2010). Part of an RNA transcript of the CRISPR locus binds to a DNase, such as Cas9, and directs it to cleave the DNA at the position homologous to the RNA, itself homologous to the inserted ("foreign")

DNA. These observations led to new methods for genetic engineering, such as the insertion or deletion of DNA or change of a gene's nucleotide sequence, at virtually any predetermined location by providing an appropriate RNA to activate the cleavage enzyme, usually Cas9 (Doudna & Charpentier, 2014; Lander, 2016).

Acquisition of new DNA spacers into the CRISPR locus requires the Cas1–Cas2 protein complex, which binds ds DNA ~33 nucleotides long and can insert it into the CRISPR locus (Nuñez, Lee, Engelman, & Doudna, 2015). In cells, an additional repeat is generated concurrently. The mechanism by which this ds DNA substrate is generated is not clear, but in *E. coli* it (or another step leading to insertion) is regulated by RecBCD and Chi. During a 16-h induction of Cas1-Cas2, *E. coli* (i.e., self) DNA spacers are inserted into the CRISPR locus (Levy et al., 2015). (The strain used lacked the destructive DNase Cas3, so cells with self DNA at the CRISPR locus survived.) Deep sequencing showed that spacer acquisition is reduced by ~30% to the 5' side of Chi (5' GCTGGTGG 3'), the side to which Chi stimulates recombination (Fig. 1); as expected, no Chi effect is observed in *recBCD* deletion mutants. These data indicate that Chi reduces acquisition of self DNA. Curiously, however, the frequency of acquisition is reduced by ~50–70% in *recBCD* deletion mutants, indicating that RecBCD is actively involved in acquisition, although RecBCD nuclease activity is reported not to be required (Radovcic et al., 2018). In addition, acquisition to the 3' side of Chi, and between Chi and an artificially introduced ds DNA break, indicates that DNA encountered by RecBCD before Chi is *not* destroyed but, rather, is actively inserted into the CRISPR locus. One possibility is that ss DNA not bound by RecA is an important precursor to CRISPR insertions. If so, acquisition to the 3' side of Chi, where RecBCD makes ss DNA, but not to the 5' side of Chi, where RecBCD loads RecA, is accounted for.

Notably, the results of Levy et al. (2015) show that self DNA is acquired; i.e., *E. coli* DNA is inserted into the *E. coli* CRISPR locus, at least in the absence of destructive Cas proteins. Presumably, these bacteria commit suicide when, as in wild-type cells, they contain the destructive Cas proteins, which would be directed to cleave the *E. coli* genome locus from which the newly acquired DNA came. These bacteria would be removed from the population, leaving only bacteria in which non-self DNA had been acquired. This scenario seems feasible, because the frequency of acquisition appears to be quite low—perhaps one bacterium in 10,000 inserts new DNA into the CRISPR locus per cell generation. If phage DNA were inserted into the CRISPR locus, these cells, though rare, would survive subsequent infection by that phage, whereas those that did not insert that phage DNA into the CRISPR locus would be killed by the phage. Eventually, only bacteria with the "correct" insertions into the CRISPR locus would survive, and the population would survive subsequent phage infection. Thus, the apparent preference for non-self DNA being in the CRISPR locus could reflect random insertion followed by self destruction, a scenario opposite to that usually considered for CRISPR.

Another type of DNA island includes retrons, which encode a bacterial reverse transcriptase (RT) (Simon, Ellington, & Finkelstein, 2019). One recently studied retron, Ec48, engages RecBCD for its action—inducing a phage-infected cell to commit suicide before production of mature phage seemingly to spare the rest of the bacterial population from infection and death (Millman et al., 2020). The Ec48 retron-encoded RT makes a 48-nucleotide DNA copy of part of the mRNA synthesized from the retron locus (Mao, Inouye, & Inouye, 1997). The 2' OH of a highly conserved G nucleotide in the RNA primes the DNA synthesis, resulting in covalent linkage of DNA and RNA. Host RNase H removes most of the RNA, leaving a DNA-RNA complex of ~170 nucleotides. When a phage that encodes a RecBCD inhibitor, such as lambda Gam, infects the cell, the inhibited RecBCD and the DNA-RNA complex activate an "effector" protein, also encoded by the retron, to kill the cell (Millman et al., 2020). The Ec48 effector protein contains two putative transmembrane regions, suggesting that activation of this protein makes the cell membrane leaky and leads to death. Although Chi has no known role in Ec48 action, RecBCD is clearly involved, in an unknown manner, in telling the cell that non-self DNA has entered the cell and it's time to commit suicide and thus protect the population.

10. Conclusion: Telling self from non-self is complex

Given that the basic genetic material in all cells and plasmids and many phages is the same—DNA—it is no surprise that cells have a hard time telling self from non-self on the basis of DNA. Even in cases where bacteria have evolved this discrimination, many phages have won the "arms race" by evolving anti-restriction and anti-CRISPR mechanisms (Stanley & Maxwell, 2018; Tock & Dryden, 2005). From this viewpoint, it is understandable that something as simple as Chi does not serve the purpose. Nevertheless, the stimulation of a critical function—DNA break repair and genetic recombination—by Chi appears widely important for cells, because the Chi-RecBCD combination appears universal in enteric bacteria, and related cohorts may exist in many other bacterial species. The intersection of RecBCD with recombination, CRISPR spacer acquisition, and retrons may reflect RecBCD's central role in maintaining life.

11. Methods of data analysis

Reference numbers for genome sequences are in Tables S3 and S11 [Supplementary Material in the online version at https://doi.org/10.1016/bs.adgen.2022.06.001]. Scripting for computational analysis of genomic data was done chiefly in Python 3.9. Scripts are available at https://drive.google.com/file/d/1ELWLMNd-s82lNUqdRKPsrbekjwLYuqQe/view?usp=sharing

Statistical tests and linear regression analyses used Graphpad and Excel.

Acknowledgments

We are grateful to Sue Amundsen, Graham Hatfull, Eduardo Rocha, Lyle Simmons, and especially Sherwood Casjens and Rasi Subramaniam for helpful discussions; and Sue Amundsen, Bob Blumenthal, Justin Courcelle, Gareth Cromie, Kevin Forsberg, Khoi Ha, Rasi Subramaniam, Yihua Zhu, and anonymous reviewers for helpful comments on the manuscript. This research was supported by MIRA grant R35 GM118120 from the National Institute of General Medical Sciences of the United States of America to GRS.

Author contributions: SS performed genome analyses, made the figures and tables, and commented on the manuscript; GRS posed questions and wrote the first draft and revisions of the manuscript.

References

Amundsen, S. K., Sharp, J. W., & Smith, G. R. (2016). RecBCD enzyme "Chi recognition" mutants recognize Chi recombination hotspots in the right DNA context. *Genetics*, *204*, 139–152.

Amundsen, S. K., & Smith, G. R. (2007). Chi hotspot activity in *Escherichia coli* without RecBCD exonuclease activity: Implications for the mechanism of recombination. *Genetics*, *175*, 41–54.

Amundsen, S. K., Taylor, A. F., Chaudhury, A. M., & Smith, G. R. (1986). *recD*: The gene for an essential third subunit of exonuclease V. *Proceedings of the National Academy of Sciences of the United States of America*, *83*, 5558–5562.

Anderson, D. G., & Kowalczykowski, S. C. (1997a). The recombination hot spot χ is a regulatory element that switches the polarity of DNA degradation by the RecBCD enzyme. *Genes & Development*, *11*, 571–581.

Anderson, D. G., & Kowalczykowski, S. C. (1997b). The translocating RecBCD enzyme stimulates recombination by directing RecA protein onto ssDNA in a χ regulated manner. *Cell*, *90*, 77–86.

Anderson, D. G., & Kowalczykowski, S. C. (1998). Reconstitution of an SOS response pathway: Derepression of transcription in response to DNA breaks. *Cell*, *95*, 975–979.

Barrangou, R., Fremaux, C., Deveau, H., Richards, M., Boyaval, P., Moineau, S., et al. (2007). CRISPR provides acquired resistance against viruses in prokaryotes. *Science*, *315*, 1709–1712.

Bell, S. J., Chow, Y. C., Ho, J. Y., & Forsdyke, D. R. (1998). Correlation of Chi orientation with transcription indicates a fundamental relationship between recombination and transcription. *Gene*, *216*, 285–292.

Biaudet, V., El Karoui, M., & Gruss, A. (1998). Codon usage can explain GT-rich islands surrounding Chi sites on the *Escherichia coli* genome. *Molecular Microbiology*, *29*, 666–669.

Biek, D. P., & Cohen, S. N. (1986). Identification and characterization of *recD*, a gene affecting plasmid maintenance and recombination in *Escherichia coli*. *Journal of Bacteriology*, *167*, 594–603.

Biswas, I., Maguin, E., Ehrlich, S. D., & Gruss, A. (1995). A 7-base-pair sequence protects DNA from exonucleolytic degradation in *Lactococcus lactis*. *Proceedings of the National Academy of Sciences of the United States of America*, *92*, 2244–2248.

Bobay, L. M., Touchon, M., & Rocha, E. P. (2013). Manipulating or superseding host recombination functions: A dilemma that shapes phage evolvability. *PLoS Genetics*, *9*, e1003825.

Bolotin, A., Quinquis, B., Sorokin, A., & Ehrlich, S. D. (2005). Clustered regularly interspaced short palindrome repeats (CRISPRs) have spacers of extrachromosomal origin. *Microbiology (Reading)*, *151*, 2551–2561.

Boycheva, S., Chkodrov, G., & Ivanov, I. (2003). Codon pairs in the genome of *Escherichia coli*. *Bioinformatics*, *19*, 987–998.

Brewer, B. J. (1988). When polymerases collide: replication and the transcriptional organization of the *Escherichia coli* chromosome. *Cell*, *53*, 679–686.

Buton, A., & Bobay, L. M. (2021). Evolution of Chi motifs in Proteobacteria. *G3 (Bethesda)*, *11*, jkaa054.

Chedin, F., Ehrlich, S. D., & Kowalczykowski, S. C. (2000). The *Bacillus subtilis* AddAB helicase/nuclease is regulated by its cognate Chi sequence *in vitro*. *Journal of Molecular Biology*, *298*, 7–20.

Chedin, F., Noirot, P., Biaudet, V., & Ehrlich, S. D. (1998). A five-nucleotide sequence protects DNA from exonucleolytic degradation by AddAB, the RecBCD analogue of *Bacillus subtilis*. *Molecular Microbiology*, *29*, 1369–1377.

Cheng, K., Wilkinson, M., Chaban, Y., & Wigley, D. B. (2020). A conformational switch in response to Chi converts RecBCD from phage destruction to DNA repair. *Nature Structural & Molecular Biology*, *27*, 71–77.

Clark, A. J., Chamberlin, M., Boyce, R. P., & Howard-Flanders, P. (1966). Abnormal metabolic response to ultraviolet light of a recombination deficient mutant of *Escherichia coli* K12. *Journal of Molecular Biology*, *19*, 442–454.

Colbert, T., Taylor, A. F., & Smith, G. R. (1998). Genomics, Chi sites and codons: "Islands of preferred DNA pairing" are oceans of ORFs. *Trends in Genetics*, *14*, 485–488.

Cromie, G. A. (2009). Phylogenetic ubiquity and shuffling of the bacterial RecBCD and AddAB recombination complexes. *Journal of Bacteriology*, *191*, 5076–5084.

Dabert, P., Ehrlich, S. D., & Gruss, A. (1992). χ sequence protects against RecBCD degradation of DNA in vivo. *Proceedings of the National Academy of Sciences of the United States of America*, *89*, 12073–12077.

Dixon, D. A., & Kowalczykowski, S. C. (1993). The recombination hotspot χ is a regulatory sequence that acts by attenuating the nuclease activity of the *E. coli* RecBCD enzyme. *Cell*, *73*, 87–96.

Don, R. H., & Pemberton, J. M. (1981). Properties of six pesticide degradation plasmids isolated from *Alcaligenes paradoxus* and *Alcaligenes eutrophus*. *Journal of Bacteriology*, *145*, 681–686.

Doudna, J. A., & Charpentier, E. (2014). Genome editing. The new frontier of genome engineering with CRISPR-Cas9. *Science, 346*, 1258096.

El Karoui, M., Biaudet, V., Schbath, S., & Gruss, A. (1999). Characteristics of Chi distribution on different bacterial genomes. *Research in Microbiology, 150*, 579–587.

El Karoui, M., Ehrlich, D., & Gruss, A. (1998). Identification of the lactococcal exonuclease/recombinase and its modulation by the putative Chi sequence. *Proceedings of the National Academy of Sciences of the United States of America, 95*, 626–631.

Ennis, D. G., Amundsen, S. K., & Smith, G. R. (1987). Genetic functions promoting homologous recombination in *Escherichia coli*: A study of inversions in phage λ. *Genetics, 115*, 11–24.

Faulds, D., Dower, N., Stahl, M. M., & Stahl, F. W. (1979). Orientation-dependent recombination hotspot activity in bacteriophage λ. *Journal of Molecular Biology, 131*, 681–695.

Garneau, J. E., Dupuis, M. E., Villion, M., Romero, D. A., Barrangou, R., Boyaval, P., et al. (2010). The CRISPR/Cas bacterial immune system cleaves bacteriophage and plasmid DNA. *Nature, 468*, 67–71.

Gilcrease, E. B., & Casjens, S. R. (2018). The genome sequence of *Escherichia coli* tailed phage D6 and the diversity of Enterobacteriales circular plasmid prophages. *Virology, 515*, 203–214.

Gurung, D., & Blumenthal, R. M. (2020). Distribution of RecBCD and AddAB recombination-associated genes among bacteria in 33 phyla. *Microbiology (Reading), 166*, 1047–1064.

Gutman, G. A., & Hatfield, G. W. (1989). Nonrandom utilization of codon pairs in *Escherichia coli. Proceedings of the National Academy of Sciences of the United States of America, 86*, 3699–3703.

Hacker, J., & Kaper, J. B. (2000). Pathogenicity islands and the evolution of microbes. *Annual Review of Microbiology, 54*, 641–679.

Halpern, D., Chiapello, H., Schbath, S., Robin, S., Hennequet-Antier, C., Gruss, A., et al. (2007). Identification of DNA motifs implicated in maintenance of bacterial core genomes by predictive modeling. *PLoS Genetics, 3*, 1614–1621.

Henderson, D., & Weil, J. (1975). Recombination-deficient deletions in bacteriophage lambda and their interaction with Chi mutations. *Genetics, 79*, 143–174.

Hille, F., Richter, H., Wong, S. P., Bratovic, M., Ressel, S., & Charpentier, E. (2018). The biology of CRISPR-Cas: Backward and forward. *Cell, 172*, 1239–1259.

Iida, S., & Arber, W. (1979). Multiple physical differences in the genome structure of functionally related bacteriophages P1 and P7. *Molecular & General Genetics, 173*, 249–261.

Ikeda, H., & Tomizawa, J.-I. (1968). Prophage P1, an extrachromosomal replication unit. *Cold Spring Harbor Symposia on Quantitative Biology, 33*, 791–798.

Ishino, Y., Krupovic, M., & Forterre, P. (2018). History of CRISPR-Cas from encounter with a mysterious repeated sequence to genome editing technology. *Journal of Bacteriology, 200*, e00580-17.

Ishino, Y., Shinagawa, H., Makino, K., Amemura, M., & Nakata, A. (1987). Nucleotide sequence of the *iap* gene, responsible for alkaline phosphatase isozyme conversion in *Escherichia coli*, and identification of the gene product. *Journal of Bacteriology, 169*, 5429–5433.

Johnson, C. M., & Grossman, A. D. (2015). Integrative and conjugative elements (ICEs): What they do and how they work. *Annual Review of Genetics, 49*, 577–601.

Kobayashi, I. (1998). Selfishness and death: Raison d'être of restriction, recombination and mitochondria. *Trends in Genetics, 14*, 368–374.

Kobayashi, I., Stahl, M. M., Fairfield, F. R., & Stahl, F. W. (1984). Coupling with packaging explains apparent nonreciprocality of Chi-stimulated recombination of bacteriophage lambda by RecA and RecBC functions. *Genetics, 108*, 773–794.

Köppen, A., Krobitsch, S., Thoms, B., & Wackernagel, W. (1995). Interaction with the recombination hot spot χ *in vivo* converts the RecBCD enzyme of *Escherichia coli* into a χ-independent recombinase by inactivation of the RecD subunit. *Proceedings of the National Academy of Sciences of the United States of America*, *92*, 6249–6253.

Kuzminov, A. (1995). Collapse and repair of replication forks in *Escherichia coli*. *Molecular Microbiology*, *16*, 373–384.

Kuzminov, A., Schabtach, E., & Stahl, F. W. (1994). χ sites in combination with RecA protein increase the survival of linear DNA in *Escherichia coli* by inactivating exoV activity of RecBCD nuclease. *EMBO Journal*, *13*, 2764–2776.

Lam, S. T., Stahl, M. M., McMilin, K. D., & Stahl, F. W. (1974). Rec-mediated recombinational hot spot activity in bacteriophage lambda. II. A mutation which causes hot spot activity. *Genetics*, *77*, 425–433.

Lander, E. S. (2016). The heroes of CRISPR. *Cell*, *164*, 18–28.

Levy, A., Goren, M. G., Yosef, I., Auster, O., Manor, M., Amitai, G., et al. (2015). CRISPR adaptation biases explain preference for acquisition of foreign DNA. *Nature*, *520*, 505–510.

Luria, S. E. (1953). Host-induced modifications of viruses. *Cold Spring Harbor Symposia on Quantitative Biology*, *18*, 237–244.

Mao, J. R., Inouye, S., & Inouye, M. (1997). msDNA-Ec48, the smallest multicopy single-stranded DNA from *Escherichia coli*. *Journal of Bacteriology*, *179*, 7865–7868.

Martinez-Martinez, L., Pascual, A., & Jacoby, G. A. (1998). Quinolone resistance from a transferable plasmid. *Lancet*, *351*, 797–799.

McKittrick, N. H., & Smith, G. R. (1989). Activation of Chi recombinational hotspots by RecBCD-like enzymes from enteric bacteria. *Journal of Molecular Biology*, *210*, 485–495.

Millman, A., Bernheim, A., Stokar-Avihail, A., Fedorenko, T., Voichek, M., Leavitt, A., et al. (2020). Bacterial retrons function in anti-phage defense. *Cell*, *183*, 1551–1561, (e1512).

Milo, R. (2013). What is the total number of protein molecules per cell volume? A call to rethink some published values. *BioEssays*, *35*, 1050–1055.

Mojica, F. J., Diez-Villasenor, C., Garcia-Martinez, J., & Soria, E. (2005). Intervening sequences of regularly spaced prokaryotic repeats derive from foreign genetic elements. *Journal of Molecular Evolution*, *60*, 174–182.

Murphy, K. C. (2000). Bacteriophage P22 Abc2 protein binds to RecC increases the 5' strand nicking activity of RecBCD and together with lambda bet, promotes Chi-independent recombination. *Journal of Molecular Biology*, *296*, 385–401.

Murray, N. E. (2002). 2001 Fred Griffith review lecture. Immigration control of DNA in bacteria: self versus non-self. *Microbiology (Reading)*, *148*, 3–20.

Myers, R. S., Kuzminov, A., & Stahl, F. W. (1995). The recombination hot spot χ activates RecBCD recombination by converting *Escherichia coli* to a *recD* mutant phenocopy. *Proceedings of the National Academy of Sciences of the United States of America*, *92*, 6244–6248.

Nuñez, J. K., Lee, A. S., Engelman, A., & Doudna, J. A. (2015). Integrase-mediated spacer acquisition during CRISPR-Cas adaptive immunity. *Nature*, *519*, 193–198.

Oliver, D. B., & Goldberg, E. B. (1977). Protection of parental T4 DNA from a restriction exonuclease by the product of gene *2*. *Journal of Molecular Biology*, *116*, 877–881.

Pavankumar, T. L., Sinha, A. K., & Ray, M. K. (2018). Biochemical characterization of RecBCD enzyme from an Antarctic Pseudomonas species and identification of its cognate Chi (χ) sequence. *PLoS One*, *13*, e0197476.

Ponticelli, A. S., Schultz, D. W., Taylor, A. F., & Smith, G. R. (1985). Chi-dependent DNA strand cleavage by RecBC enzyme. *Cell*, *41*, 145–151.

Pourcel, C., Salvignol, G., & Vergnaud, G. (2005). CRISPR elements in *Yersinia pestis* acquire new repeats by preferential uptake of bacteriophage DNA, and provide additional tools for evolutionary studies. *Microbiology (Reading)*, *151*, 653–663.

Radovcic, M., Killelea, T., Savitskaya, E., Wettstein, L., Bolt, E. L., & Ivancic-Bace, I. (2018). CRISPR-Cas adaptation in *Escherichia coli* requires RecBCD helicase but not nuclease activity, is independent of homologous recombination, and is antagonized by 5' ssDNA exonucleases. *Nucleic Acids Research*, *46*, 10173–10183.

Rao, B. J., Chiu, S. K., Bazemore, L. R., Reddy, G., & Radding, C. M. (1995). How specific is the first recognition step of homologous recombination? *Trends in Biochemical Sciences*, *20*, 109–113.

Rocha, E. P. C. (2004). The replication-related organization of bacterial genomes. *Microbiology (Reading)*, *150*, 1609–1627.

Rosamond, J., Telander, K. M., & Linn, S. (1979). Modulation of the *RecBC* enzyme of *Escherichia coli* K12 by Ca^{2+}. *Journal of Biological Chemistry*, *254*, 8646–8652.

Rostol, J. T., & Marraffini, L. (2019). (Ph)ighting phages: How bacteria resist their parasites. *Cell Host & Microbe*, *25*, 184–194.

Sarthy, P. V., & Meselson, M. (1976). Single burst study of rec- and red-mediated recombination in bacteriophage lambda. *Proceedings of the National Academy of Sciences of the United States of America*, *73*, 4613–4617.

Schroeder, J. W., Sankar, T. S., Wang, J. D., & Simmons, L. A. (2020). The roles of replication-transcription conflict in mutagenesis and evolution of genome organization. *PLoS Genetics*, *16*, e1008987.

Schultz, D. W., & Smith, G. R. (1986). Conservation of Chi cutting activity in terrestrial and marine enteric bacteria. *Journal of Molecular Biology*, *189*, 585–595.

Shepherd, J. C. (1981). Periodic correlations in DNA sequences and evidence suggesting their evolutionary origin in a comma-less genetic code. *Journal of Molecular Evolution*, *17*, 94–102.

Simmon, V. F., & Lederberg, S. (1972). Degradation of bacteriophage lambda deoxyribonucleic acid after restriction by *Escherichia coli* K-12. *Journal of Bacteriology*, *112*, 161–169.

Simon, A. J., Ellington, A. D., & Finkelstein, I. J. (2019). Retrons and their applications in genome engineering. *Nucleic Acids Research*, *47*, 11007–11019.

Smith, G. R. 1991. Conjugational recombination in E. coli: Myths and mechanisms. *Cell 64*, 19–27.

Smith, G. R. (2012). How RecBCD and Chi promote DNA break repair and recombination—A molecular biologist's view. *Microbiology and Molecular Biology Reviews*, *76*, 217–228.

Smith, G. R., Roberts, C. M., & Schultz, D. W. (1986). Activity of Chi recombinational hotspots in *Salmonella typhimurium*. *Genetics*, *112*, 429–439.

Sourice, S., Biaudet, V., El Karoui, M., Ehrlich, S. D., & Gruss, A. (1998). Identification of the Chi site of *Haemophilus influenzae* as several sequences related to the *Escherichia coli* Chi site. *Molecular Microbiology*, *27*, 1021–1029.

Sprague, K. U., Faulds, D. H., & Smith, G. R. (1978). A single base-pair change creates a Chi recombinational hotspot in bacteriophage λ. *Proceedings of the National Academy of Sciences of the United States of America*, *75*, 6182–6186.

Stahl, F. W. (2005). Chi: A little sequence controls a big enzyme. *Genetics*, *170*, 487–493.

Stahl, F. W., Lieb, M., & Stahl, M. M. (1984). Does Chi give or take? *Genetics*, *108*, 795–808.

Stahl, F. W., Shurvinton, C. E., Thomason, L. C., Hill, S., & Stahl, M. M. (1995). On the clustered exchanges of the RecBCD pathway operating on phage λ. *Genetics*, *139*, 1107–1121.

Stahl, F. W., & Stahl, M. M. (1977). Recombination pathway specificity of Chi. *Genetics*, *86*, 715–725.

Stahl, F. W., Stahl, M. M., Malone, R. E., & Crasemann, J. M. (1980). Directionality and nonreciprocality of Chi-stimulated recombination in phage λ. *Genetics*, *94*, 235–248.

Stahl, F. W., Stahl, M. M., Young, L., & Kobayashi, I. (1982). Chi-stimulated recombination between phage λ and the plasmid λdv. *Genetics*, *102*, 599–613.
Stahl, F. W., Thomason, L. C., Siddiqi, I., & Stahl, M. M. (1990). Further tests of a recombination model in which χ removes the RecD subunit from the RecBCD enzyme of *Escherichia coli*. *Genetics*, *126*, 519–533.
Stanley, S. Y., & Maxwell, K. L. (2018). Phage-encoded anti-CRISPR defenses. *Annual Review of Genetics*, *52*, 445–464.
Sternberg, N., & Hoess, R. (1983). The molecular genetics of bacteriophage P1. *Annual Review of Genetics*, *17*, 123–154.
Taylor, A. F., Amundsen, S. K., & Smith, G. R. (2016). Unexpected DNA context-dependence identifies a new determinant of Chi recombination hotspots. *Nucleic Acids Research*, *44*, 8216–8228.
Taylor, A. F., Schultz, D. W., Ponticelli, A. S., & Smith, G. R. (1985). RecBC enzyme nicking at Chi sites during DNA unwinding: Location and orientation dependence of the cutting. *Cell*, *41*, 153–163.
Taylor, A. F., & Smith, G. R. (1992). RecBCD enzyme is altered upon cutting DNA at a Chi recombination hotspot. *Proceedings of the National Academy of Sciences of the United States of America*, *89*, 5226–5230.
Taylor, A. F., & Smith, G. R. (1995). Strand specificity of nicking of DNA at Chi sites by RecBCD enzyme: Modulation by ATP and magnesium levels. *Journal of Biological Chemistry*, *270*, 24459–24467.
Taylor, A. F., & Smith, G. R. (1999). Regulation of homologous recombination: Chi inactivates RecBCD enzyme by disassembly of the three subunits. *Genes & Development*, *13*, 890–900.
Tock, M. R., & Dryden, D. T. (2005). The biology of restriction and anti-restriction. *Current Opinion in Microbiology*, *8*, 466–472.
Touzain, F., Petit, M. A., Schbath, S., & El Karoui, M. (2011). DNA motifs that sculpt the bacterial chromosome. *Nature Reviews. Microbiology*, *9*, 15–26.
Uno, R., Nakayama, Y., & Tomita, M. (2006). Over-representation of Chi sequences caused by di-codon increase in *Escherichia coli* K-12. *Gene*, *380*, 30–37.
Wright, M., Buttin, G., & Hurwitz, J. (1971). The isolation and characterization from *Escherichia coli* of an adenosine triphosphate-dependent deoxyribonuclease directed by *rec* B, C genes. *Journal of Biological Chemistry*, *246*, 6543–6555.
Yun, T., & Vapnek, D. (1977). Electron microscopic analysis of bacteriophages P1, P1Cm, and P7. Determination of genome sizes, sequence homology, and location of antibiotic-resistance determinants. *Virology*, *77*, 376–385.
Zabrovitz, S., Legev, N., & Cohen, G. (1977). Growth of bacteriophage P1 in recombination-deficient hosts of *Escherichia coli*. *Virology*, *80*, 233–248.

CHAPTER TWO

Multiple regulatory mechanisms for pH homeostasis in the gastric pathogen, *Helicobacter pylori*

Xuhua Xia[a,b,*]
[a]Department of Biology, University of Ottawa, Ottawa, Canada
[b]Ottawa Institute of Systems Biology, Ottawa, Canada
*Corresponding author: e-mail address: xxia@uottawa.ca

Contents

Abstract

Acid-resistance in gastric pathogen *Helicobacter pylori* requires the coordination of four essential processes to regulate urease activity. Firstly, urease expression above a base level needs to be finely tuned at different ambient pH. Secondly, as nickel is needed to activate urease, nickel homeostasis needs to be maintained by proteins that import and export nickel ions, and sequester, store and release nickel when needed. Thirdly, urease accessary proteins that activate urease activity by nickel insertion need to be expressed. Finally, a reliable source of urea needs to be maintained by both intrinsic and extrinsic sources of urea. Two-component systems (arsRS and flgRS), as well as a nickel response regulator (NikR), sense the change in pH and act on a variety of genes to accomplish the function of acid resistance without causing cellular overalkalization and nickel toxicity. Nickel storage proteins also feature built-in switches to store nickel at neutral pH and release nickel at low pH. This review summarizes the current status of *H. pylori* research and highlights a number of hypotheses that need to be tested.

Advances in Genetics, Volume 109
ISSN 0065-2660
https://doi.org/10.1016/bs.adgen.2022.07.001

1. Introduction

Helicobacter pylori colonizes mammalian stomach, causing gastric and duodenal ulcers and gastric cancer in human (Correa, 1997; Hamajima et al., 2004; Hunt, 2004; Menaker, Sharaf, & Jones, 2004; Siavoshi, Malekzadeh, Daneshmand, Smoot, & Ashktorab, 2004; Wroblewski, Peek, & Wilson, 2010). The bacterium is acid-resistant but not acidophilic because its optimal growth pH is neutral or slightly above neutral (Bauerfeind, Garner, Dunn, & Mobley, 1997; Rektorschek et al., 2000; Sachs, Meyer-Rosberg, Scott, & Melchers, 1996; Scott, Marcus, Weeks, & Sachs, 2002). The gastric fluid in human stomach has a mean pH of about 1.4 over a 24-h period (Sachs, Weeks, Melchers, & Scott, 2003; Savarino et al., 1996). Because most of *H. pylori* reside in the gastric mucus near the epithelial cell surface (Hessey et al., 1990; Schreiber et al., 2004; Stingl & De Reuse, 2005; Thomsen, Gavin, & Tasman-Jones, 1990), it was once hypothesized that *H. pylori* might invade mammalian stomach with a meal which transiently increases the stomach pH (Dressman et al., 1990; Ovesen et al., 1986; Russell et al., 1993), so that some of the bacteria would find their way into the less acidic mucous gel layer in the less acidic part of the stomach before the pH in the stomach fluid decreases to the pre-meal level. Such a hypothesis would require little acid-resistance mechanism, and was rejected for the following reasons. Although the buffering action of the gastric epithelium and limited acid diffusion through the gastric mucus could alleviate the hostile environment of mammalian stomach, both empirical studies (Allen, Flemstrom, Garner, & Kivilaakso, 1993) and theoretical modeling (Engel, Peskoff, Kauffman, & Grossman, 1984) have suggested that the mucus protection is rather limited (Matin, Zychlinsky, Keyhan, & Sachs, 1996; Sachs et al., 2003). In particular, a trans-mucus pH gradient is abolished when the luminal pH drops to <2.5 (Baumgartner & Montrose, 2004). It is therefore necessary for *H. pylori* to have acid-resistance mechanisms to overcome the acidic conditions in vitro (Sachs et al., 2003) or in vivo (Debowski et al., 2017).

1.1 Two mechanisms of acid-resistance

H. pylori features two mechanisms for acid-resistance. The first is a short-term mechanism mediated by a positively charged membrane to alleviate the influx of protons (Alm et al., 2000; Huynen, Dandekar, & Bork, 1998; Sachs et al., 2003; Scott, Weeks, Melchers, & Sachs, 1998; Solnick, Hansen, Salama, Boonjakuakul, & Syvanen, 2004; Tomb et al., 1997; Valenzuela, Cerda, & Toledo, 2003; Yamaoka et al., 2002). A recent study of 34 membrane proteins of *H. pylori* STR 26695 (Baik et al., 2004) confirmed high protein isoelectric point (pI) in membrane proteins in *H. pylori*. Membrane proteins in *H. pylori* have pI significantly higher than those in *H. hepaticus* (Xia, 2018a; Xia & Palidwor, *2005*). The high membrane protein pI implies a strongly positively charged shell, which would alleviate the influx of proton, buying time for *H. pylori* to find a less acidic niche inside mammalian stomach, and to allow the second and long-term acid-resistant mechanism to function.

The second, and much more thoroughly studied, acid-resistance mechanism in *H. pylori* is conferred by the urease gene cluster *ureABIEFGD* (where D was designated by H in early literature). Urease accounts for ~10% of total protein content of *H. pylori* at pH 6–7 (Bauerfeind et al., 1997). It is a multi-unit enzyme encoded by *ureAB*. The UreA and UreB are assembled into a dimer UreAB which trimerizes to form $(UreAB)_3$ which, in turn, are finally assembled into a tetramer, $[(UreAB)_3]_4$ (Carter, Flugga, Boer, Mulrooney, & Hausinger, 2009; Ha et al., 2001). The urease, once activated through nickel insertion, converts urea to two molecules of ammonia and a carbonic acid, with the net effect of an increased pH (Mobley, Hu, & Foxal, 1991; Mobley, Island, & Hausinger, 1995; Rektorschek et al., 2000; Sachs et al., 2003; Stingl, Altendorf, & Bakker, 2002). The cytoplasmic pH of *H. pylori* is increased only in the presence of urea and urease activity but not in urease-negative strains (Clyne, Labigne, & Drumm, 1995). This urease-mediated pH buffering allows *H. pylori* to survive for at least 3h at pH 1 with urea (Stingl et al., 2001) and to maintain a nearly neutral cytoplasmic pH between 5.0 and 7.0 under gastric acidity (Matin et al., 1996; Scott et al., 2002; Stingl, Uhlemann, Schmid, Altendorf, & Bakker, 2002).

The essential function of the urease gene cluster in mediating pH homeostasis in *H. pylori* was challenged by the observation that urease-negative strains of *H. pylori* was observed to colonize gerbil stomach and induce ulcer (Mine, Muraoka, Saika, & Kobayashi, 2005). However, it was known that gerbil stomach is less acidic than stomach in human (Sachs et al., 2003), monkey (Kararli, 1995) and carrion-eaters and carnivores (Beasley, Koltz, Lambert, Fierer, & Dunn, 2015). Urease-negative

H. pylori could survive in pH as low as 4.5 (Clyne et al., 1995), and may therefore survive and reproduce in the less acidic gerbil stomach. Many other studies (Clyne et al., 1995; Eaton, Brooks, Morgan, & Krakowka, 1991; Eaton & Krakowka, 1994; Mobley et al., 1995; Takahashi et al., 1993; Tsuda, Karita, Morshed, Okita, & Nakazawa, 1994), have demonstrated the essential requirement of urease and urea for successful gastric colonization by *H. pylori*.

1.2 Urease activity needs to be regulated

Urease activity needs to be regulated for two reasons. First, urease activity is useful only under acidic environment such as in mammalian stomach. Second, catalyzing urea at neutral pH by urease in *H. pylori* would lead to cytoplasmic overalkalization and ammonia toxicity that inhibit *H. pylori* proliferation (Clyne et al., 1995; Meyer-Rosberg, Scott, Rex, Melchers, & Sachs, 1996; Mobley et al., 1995; Morgan, Freedman, Depew, & Kraft, 1987), so it makes sense for *H. pylori* cultured at near-neutral pH to lose urease activity completely (Andersen et al., 1997). Much progress has been made in the search for the regulatory mechanism of urease in *H. pylori* in the last 20 years, which I aim to summarize in this review in the hope of highlighting what is missing and pointing out future directions.

Theoretically, there could be many possible cellular regulatory mechanisms. Any regulatory mechanism would need a minimum of a sensor that senses the environmental change and an effector that allows the production and activation of one or more products (typically proteins) in response to the changed environment. This regulation can occur at the translation level or at the transcription level, even within a single protein or within a single amino acid. For example, histidine residues have pK_a values between 6.0 and 6.5 and is not protonated at neutral pH. A decrease of pH below 6 will result in positive charge of histidine residues. A histidine-rich protein may bind to Ni^{2+} at neutral pH, but will tend to release Ni^{2+} at low pH when the protein become strongly positively charged. A histidine residue (H193) is also used as a pH sensor in UreI that changes conformation at low pH to facilitate import of urea into the *H. pylori* cell (Bury-Moné, Skouloubris, Labigne, & De Reuse, 2001). A single mutation at H193 is sufficient to abolish UreI's response to reduced pH (Bury-Moné et al., 2001).

Most of the regulatory mechanisms in response to environmental change in bacteria work at the transcription level through two-component systems (TCSs) (Miguel-Romero, Mideros-Mora, Marina, & Casino, 2020). Here a histidine residue again serves as a key component of a pH sensor. A bacterial TCS, in its most basic form, consists of two protein partners: a membrane sensor protein (S) that is a histidine kinase, and a cytoplasmic response protein (R) that is typically a transcription factor serving as the effector. A TCS works by two chemical reactions involving phosphate groups (Capra & Laub, 2012; Sanders, Gillece-Castro, Burlingame, & Koshland, 1992; Szurmant & Hoch, 2010). In the first reaction, the S member of the TCS, upon sensing a specific change in the environment (e.g., pH or nutrient availability) that may perturb homeostasis, undergoes ATP-dependent autophosphorylation of a histidine residue. In the second reaction, the S partner donates the phosphate group to a conserved aspartate residue in the R partner of the TCS (Lewis, Brannigan, Muchová, Barák, & Wilkinson, 1999; Miguel-Romero et al., 2020). The R partner will then bind to the regulatory motifs in DNA to increase or suppress the expression of relevant genes as a response to the environmental change to restore homeostasis. Once the environmental change disappears, the phosphatase function of the S partner removes the phosphate group in the aspartate residue in the R member to restore the cell to the pre-change state. Regulation of urease activities involves at least two TCSs (arsRS and flgRS) which I will review in detail later.

2. *H. pylori* reference genome, gene nomenclature and genomic diversity

This section may not be needed for veteran researchers in *H. pylori* but essential for newcomers because of two sources of confusion in gene regulation literature related to *H. pylori*. The first source of confusion arose from sequencing and annotation errors and consequent removal and replacement of reference genomes for *H. pylori* in NCBI. The reference genome for *H. pylori* has been replaced at least twice, first from NC_000915 (strain 26695) to NC_017379 (strain Puno135), and then to NZ_CP071982 (strain MT5135). Many early studies on *H. pylori* are based on the strain 26695 with gene designation in NC_00915 which has long been deleted from NCBI's Genome database. This results in many orphaned gene designations in *H. pylori* strain 26695. For example, Stähler et al. (2006) documented an RND transporter in *H. pylori* strain 26695 that pumps surplus cadmium, zinc and nickel ions out of the cell. The three component genes were annotated in the original *H. pylori* reference genome (NC_000915) with locus_tags HP0969, HP0970 and HP0971, respectively. Stähler et al. (Stähler et al., 2006) named these three genes *cznA*, *cnzB* and *cnzC*, respectively. However, because these gene names were not accepted by NCBI curating staff, they were not used in GenBank records. After the reference genome (NC_000915) became obsolete, purged from GenBank database, and replaced with a new genome record, the three locus_tags were changed to HP_RS04740, HP_RS04745 and HP_RS04750, respectively. The current GenBank record of *H. pylori* strain 26695 is NC_018939, with the three locus_tags again changed to HP1CH_RS04960, HP1CH_RS04965, and HP1CH_RS04970, respectively. If one searched NCBI databases using gene names *cznA*, *cnzB* and *cnzC* or locus_tags HP0969, HP0970 and HP0971 used in the paper, one would find no trace of them. These three genes in the *H. pylori* reference genome (NZ_CP071982) have locus_tags J5D21_RS04645, J5D21_RS04650 and J5D21_RS04655, respectively.

One reason for the replacement of reference genomes for *H. pylori* is due to annotation errors in earlier reference genomes. For example, the key transcription regulator for urease activity is the two-component system arsRS, elucidated mostly by experiments with the *H. pylori* strain 26695. Unfortunately, the original reference genome for strain 26695 (NC_000915), as well as the current RefSeq genome (NC_018939) annotated ArsS incorrectly. The coding sequence (CDS) should have ended at the first stop codon UGA, but the annotated CDS included 29 extra codons. This leads to four inframe stop codons at the 3' end, and the gene is consequently (and wrongly) marked as a pseudogene. Numerous early studies demonstrated the pH-sensing function of this "pseudogene". Another obvious error involves the paralogues of *ureAB* in NC_017379 (strain Puno135). The additional paralog of *ureB* (HPPN135_RS05270, at sites 1108810..1110513) was misannotated as "urease subunit alpha," whereas the paralogous *ureA* (HPPN135_RS05265, at sites 1108110..1108793) was misannotated as "urease subunit beta."

Different *H. pylori* strains are genetically diverse (Fig. 1) and do not share all genes relevant for acid-resistance. For example, the current reference genome (NZ_CP071982) based on the genomic sequence of *H. pylori* strain MT5135 does not have the important *hpn* gene encoding the nickel-sequestrating Hpn protein that is present in strains 26695, J99, B128, and Puno135. Similarly, *H. pylori* Puno135 (NC_017379) contains paralogues of *ureAB* whereas strains MT5135 (NZ_CP071982), J99 (NC_000921) and 26695 (NC_000915) do not. A gene present in two strains may differ in sequence. For example, comparing UreF between strains MT5135 and

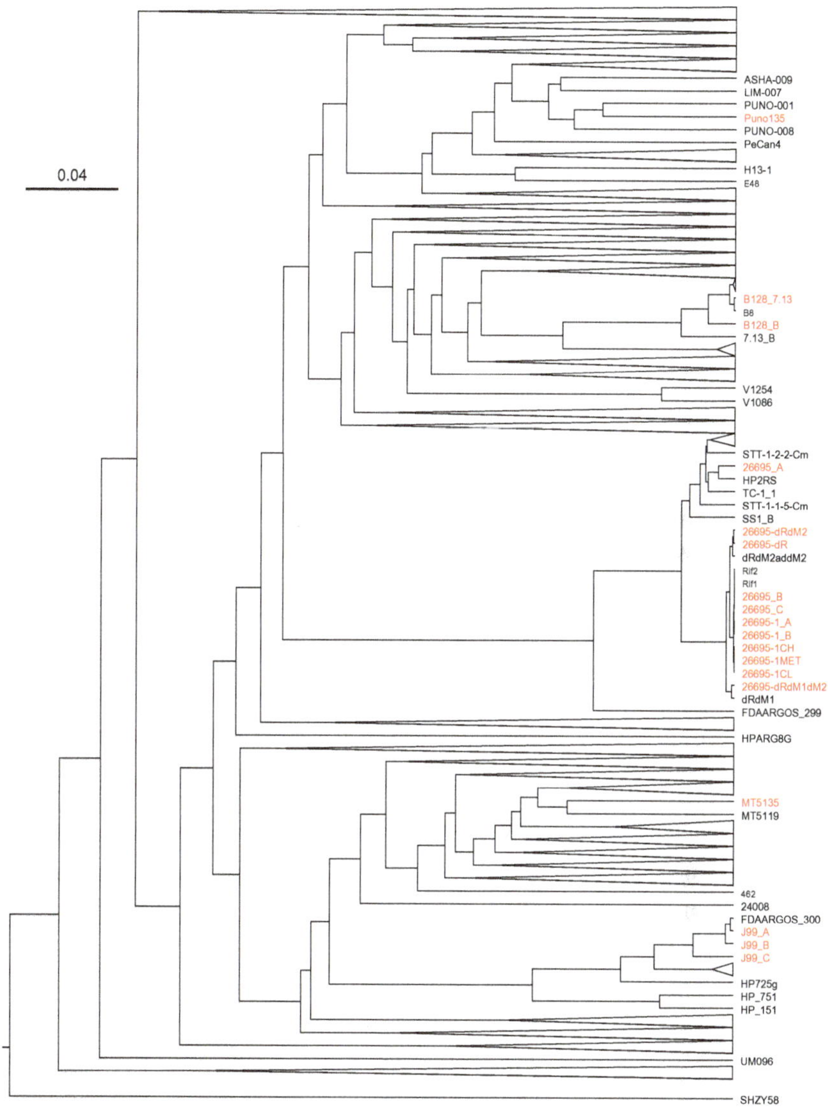

Fig. 1 Genetic diversity of *H. pylori* strains, based on 1982 isolates. Strains commonly used in experiments (26695, J99, B128, Puno135, MT5135) are colored in red. Wedges represent collapsed subtrees to avoid clutter.

Puno135 shows that UreF in Puno135 has lost 16 amino acids at the N-terminus because it has lost an upstream start codon, so translation starts at a downstream AUG codon according to the annotated genome (NC_017379). This difference in start codon would lead to substantial difference in protein isoelectric point for UreF between the two strains, being 8.6 in the MT5135 strain but 6.63 in the Puno135 strain. Such a difference in pI in UreF between the two strains suggests that UreF in Puno135 can sense pH better than UreF in MT5135. UreF in the latter will remain positively charged when pH decreases from 7 to 5, but UreF in Puno135 will change from slightly negatively charged to positively charged. These differences among *H. pylori* strains may contribute to difference in acid-resistance and explain the differential distribution of different *H. pylori* strains in different gastric niches. For example, *H. pylori* strain SS1 was more abundant than strain X47 in the gastric antrum, but opposite in the corpus (Akada et al., 2003).

The genetic diversity among strains highlights the difficulty of choosing a reference genome and the importance of studying representative strains in different clades of the tree (Fig. 1). The tree was generated by aligning 16S rRNA sequences with MAFFT (Katoh, Kuma, Miyata, & Toh, 2005) with the most accurate LINSI option ("–localpair" and "–maxiterate = 1000") and phylogenetic reconstruction with PhyML v3.3 (Guindon & Gascuel, 2003). For PhyML, the GTR + Γ substitution model was used, and the tree improvement option ("-s") was set to 'BEST' (best of NNI and SPR search). The "-o" option was set to "tlr" which optimizes the topology, the branch lengths and rate parameters.

The genetic diversity among *H. pylori* genomes is reflected in both gene content, which varies from 1334 protein coding genes in strain K154 to 1720 in strain B8, and genome size which varies from 1507930 nt in strain Aklavik86 to 1745797 nt in strain LIM-010. The reference genome from strain MT5135 has 1434 protein-coding genes in a genome of 1624458 nt. While my review of acid-resistant mechanism mediated by the urease gene cluster was written as if it is a shared feature of all *H. pylori* strains, it is important to keep in mind that some components of the acid-resistant mechanism is not applicable to all *H. pylori* strains.

The second source of confusion is the nonstandard gene nomenclature (Mobley et al., 1995). The urease gene cluster in *H. pylori* contains seven genes conventionally designated as *ureABIEFGH*. This urease gene cluster is organized in two operons (*ureAB* and *ureIEFGH*) (Akada, Shirai, Takeuchi, Tsuda, & Nakazawa, 2000). Gene *ureH* has been found in different bacterial species both upstream (e.g., *Proteus mirabilis* and *Klebsiella aerogenes*) and downstream (e.g., *Yersinia enterocolitica* and *H. pylori*) of other urease genes. The gene was conventionally named *ureD* when found upstream, but *ureD* or *ureH* when found downstream, e.g., *ureD* in *Y. enterocolitica* but *ureH* in early literature of *H. pylori*. The reference genome

(NZ_CP071982), as well as the previous reference genome (NC_017379), renames *ureH* to *ureD*, so the urease gene cluster will be referred to as *ureABIEFGD* (instead of the traditional *ureABIEFGH*). Keep in mind that *ureD* and *ureH* are synonymous in *H. pylori*.

Genes *hspA* and *hspB* in *H. pylori* literature are homologous to, and annotated in the reference genome as, *groES* and *groL*, respectively, consistent with other bacterial model species such as *Escherichia coli*. Gene *nixA* in *H. pylori* literature is represented by the gene with locus_tag *J5D21_RS01795* in the *H. pylori* reference genome without an explicit *nixA* gene or product name. Genes *hpn* and *hpnL* are paralogous but *hpnL* is often named as *hpn-2* in *H. pylori* literature (de Reuse, Vinella, & Cavazza, 2013; Vinella et al., 2015).

Table 1 shows basic information of the urease gene cluster and the associated genes that participate in the regulation of urease activities. I will discuss these genes and their regulation in detail in following sections. I_{TE} in

Table 1 Basic gene information for the urease gene cluster and the associated genes regulating urease activities in the reference genome (NZ_CP071982, strain MT5135).

Gene	Locus_tag (old in 26695[a])	Length	GC%	I_{TE}[b]	pI[c]
ureD	J5D21_RS00340 (HP0067)	798	43.86	0.79366	6.17
ureG	J5D21_RS00345 (HP0068)	600	45.67	0.83087	4.85
ureF	J5D21_RS00350 (HP0069)	774	40.96	0.82753	8.60
ureE	J5D21_RS00355 (HP0070)	513	40.94	0.77386	9.08
ureI	J5D21_RS00360 (HP0071)	588	43.88	0.81396	5.82
ureB	J5D21_RS00365 (HP0072)	1710	44.56	0.85136	5.93
ureA	J5D21_RS00370 (HP0073)	717	42.12	0.84879	9.00
arsS	J5D21_RS00795 (HP0165)	1284	36.06	0.81696	7.00
arsR	J5D21_RS00800 (HP0166)	678	39.09	0.83546	5.18
flgS	J5D21_RS01185 (HP0244)	1161	36.95	0.82367	8.86
hypA	J5D21_RS04150 (HP0869)	354	39.55	0.84827	5.35
hypB	J5D21_RS04320 (HP0900)	729	41.15	0.8162	5.57
rocF	J5D21_RS07275 (HP1399)	969	40.56	0.81609	6.05
nikR	J5D21_RS06380 (HP1338)	447	40.94	0.84587	5.76
groES	J5D21_RS00070 (HP0011)	357	40.90	0.83058	6.59

Continued

Table 1 Basic gene information for the urease gene cluster and the associated genes regulating urease activities in the reference genome (NZ_CP071982, strain MT5135).—cont'd

Gene	Locus_tag (old in 26695[a])	Length	GC%	I_{TE}[b]	pI[c]
groL	J5D21_RS00065 (HP0010)	1641	45.58	0.81458	5.37
nixA	J5D21_RS01795 (HP1077)	996	42.17	0.7936	9.46
Fur	J5D21_RS02045 (HP1027)	453	39.96	0.87335	7.81
hpnL	J5D21_RS06725 (HP_1432)	213	48.36	0.84138	7.22
hpn[d]	C528_RS07725 (HP_1427)	183	0.52	0.90187	6.82
cznA	J5D21_RS04645 (HP0969)	3060	41.83	0.79135	9.31
cznB	J5D21_RS04650 (HP0970)	1080	39.63	0.78971	8.89
cznC	J5D21_RS04655 (HP0971)	1284	39.02	0.83925	9.92

[a]Locus_tag in the obsolete genome record from strain 26695 (NC_000915), but frequently referred in literature.
[b]Index of translation efficiency (Xia, 2015) implemented in DAMBE (Xia, 2018b).
[c]Theoretical protein isoelectric point, computed with an iterative algorithm (Xia, 2018a).
[d]The *hpn* gene is absent in the reference genome, so the gene is taken from *H. pylori* strain A45 (NZ_CP053256).

Table 1 is the index of translation efficiency (Xia, 2015) implemented in DAMBE (Xia, 2018b). It is a generalization of the codon adaptation index (Sharp & Li, 1987; Xia, 2007) by accounting for background mutation bias. As most of the genes in the table are expected to respond rapidly to pH change, their mRNA should be translated efficiently and therefore should have an I_{TE} greater than an average gene. Indeed, their average I_{TE} (=0.8269) is significantly greater than the average I_{TE} for the rest of the genes (=0.8122), with $n_1=23$, $n_2=1473$, DF=1494, t=3.3053, $P<0.001$. The protein isoelectric point (pI in the last column in Table 1) is important because it is related to sensing of pH change. Take for example nickel storage protein HspA (GroES) and Hpn with pI equal to 6.59 and 6.82, respectively (Table 1). They carry little charge near pH of 7. However, when pH drops to 5, they both become positively charge and would tend to release the bound Ni^{2+}. This would make available Ni^{2+} needed for urease maturation.

3. Urease function and urea supply

The constitutively expressed cytoplasmic urease is encoded by *ureAB* which features an upstream promoter (P_{ureA}) (Akada et al., 2000; Cussac,

Ferrero, & Labigne, 1992). The evidence for the constitutive expression of urease is that it is present in large quantities even at nearly neutral pH of 6–7, accounting for ~10% of total proteins in *H. pylori* cells (Bauerfeind et al., 1997). The basal level of urease production is maintained under various conditions (Davies et al., 2002; Shirai, Fujinaga, Akada, & Nakazawa, 1999). This basal level of urease ensures the readiness of *H. pylori* to respond to acid challenge when the bacteria was swallowed into the acidic mammalian stomach.

The enzyme catalyzes urea to generate $2NH_3 + CO_2$ to buffer against the H^+ influx into either the periplasm or the cytoplasm (Mobley et al., 1991; Rektorschek et al., 2000; Sachs et al., 2003; Stingl, Altendorf, & Bakker, 2002) and to facilitate the extrusion of H^+ from the cytoplasm in the form of NH_4^+ (Stingl, Altendorf, & Bakker, 2002). However, urease is an apoenzyme requiring nickel insertion to be active. The *ureIEFGD* operon, whose expression is controlled by the promoter P_{ureI} (Akada et al., 2000; Cussac et al., 1992), encodes a membrane ureaporter (UreI) and four urease accessary proteins (UreE, UreF, UreG and UreD/ureH) that insert nickel into the urease. The expression of these genes is acid-gated (Scott et al., 2010; Weeks & Sachs, 2001; Wen, Feng, Scott, Marcus, & Sachs, 2009). The proteins from these genes, through their function of nickel insertion into urease, activates urease leading to sustained urease activity (Sachs et al., 2003; Wen et al., 2003; Williams, Preston, Hossack, Slater, & McColl, 1996). As a consequence of nickel insertion, the same amount of urease generates much higher urease activity at pH 5.5 than at pH 7 in *H. pylori* cultures (Akada et al., 2000).

The urease activity can only be sustained with a constant supply of urea as its substrate. The bacterial cell has two sources of urea supply, one internal and one external. The stomach fluid represents the external source. UreI (Bury-Moné et al., 2001; Scott et al., 2000; Weeks, Eskandari, Scott, & Sachs, 2000) is a membrane ureaporter. Its pI is 5.82, so its net electric charge will change from negative to positive when pH changed from 7 to 5 or lower. The protein changes structure upon exposure to gastric acid, which results in a large urea influx into the bacterial cell (Scott et al., 2010; Weeks & Sachs, 2001; Wen et al., 2009). The internal source of urea comes from efficient conversion of arginine to urea (and ornithine) in the cytoplasm by the highly expressed arginase (encoded by the *rocF* gene) in *H. pylori* (Hovey et al., 2007; McGee et al., 2004; Mendz & Hazell, 1996).

This internal source of urea by arginase has two implications. First, arginine is underused in *H. pylori* proteins because it is channeled to supply urea.

The positively charged membrane in *H. pylori* is mainly maintained by proteins with a surplus of positively charged lysine residues (Xia, 2018a; Xia & Palidwor, 2005). The statement "The basic amino acids, arginine and lysine, occur twice as frequently in *H. pylori* proteins as in those of *H. influenzae* and *E. coli*, perhaps reflecting an adaptation of *H. pylori* to gastric acidity" (Tomb et al., 1997) is not quite accurate. Based on all 1519 coding sequences (CDSs) in *H. pylori* Puno135 (NC_017379) and all 4140 CDSs in *E. coli* K12 MG1655 (NC_000913), arginine usage is only 3.5% in *H. pylori* but 5.5% in *E. coli*, contradicting the statement above. However, the lysine usage is 9.0% in *H. pylori* but only 4.4% in *E. coli*. The second implication of the arginase-mediated internal source of urea suggests that ureI is helpful in gaining the external source of urea, but it is not essential for the acid-resistance mechanism because of the internal source. Experimental results showed that ureI is indeed not essential for acid-resistance (Cussac et al., 1992). However, arginase activity differs substantially among different *H. pylori* strains (Hovey et al., 2007), so conceivably ureI would play a much more significant role in acid resistance in strains with weak arginase activity. The observation that UreA, UreB and UreE are recruited to UreI at the inner membrane to form a complex suggests the importance of UreI as a ureaporter, i.e., active urease should be right at the site where urea is pumped in by UreI (Scott et al., 2010).

4. Regulation of urease activities

Given the function of urease gene cluster *ureABIEFGD*, regulation of urease activity can occur at multiple levels to achieve the objective of low or no urease activity at pH 7 or above, and high urease activity at low pH. Urease activity requires the presence of all of the following eight components: UreA, B, E, F, G, D, urea and nickel. Missing anyone of the eight components would abolish urease activity. If each component is represented by the binary presence/absence states, then there are 256 different state combinations that would result in no urease activity, and only one combination (when all eight components are in "present" state) would result in urease activity. Table 2 lists a subset of these different combinations or scenarios, with only scenario 7 resulting in urease activity. Note that a reliable source of urea requires ureI and arginase encoded by *rocF*, and nickel homeostasis requires *hpnL*, *hspA*, *hspB*, *nikR* and *nixA/J5D21_01795* reviewed in detail later.

In Scenario 1 (Table 2), transcription or translation of *ureAB* is suppressed so no urease activity is possible. In Scenario 2, UreA and UreB are produced

Table 2 A subset of different combinations of proteins, urea and nickel that would achieve low or high urease activities (activity) under different pH.

Scenario	ureAB	ureEFGD	Urea	Nickel
1	−			
2	+	−		
3	+	+	−	
4	+	+	+	−
5	+	+	−	+
6	+	−	+	+
7	+	+	+	+

Presence/absence are indicated by ±.

but nickel-inserting proteins ureEFGD are not produced, so urease activity is again missing. Scenarios 1–6 represent a subset of possible scenarios when urease activity is not needed, e.g., when pH is neutral or higher. Scenario 7 (Table 2) is the only one with urease activity. It represents a desirable state under acidic pH when urease activity is needed for acid resistance. It is crucially important to safeguard against accidental activation of urease activity at neutral pH because over alkalization and ammonia toxicity can reduce the fitness of *H. pylori* (Clyne et al., 1995; Meyer-Rosberg et al., 1996; Mobley et al., 1995).

4.1 Sensors of pH changes

Any regulatory mechanism requires a sensor of environmental change. How is a reduction of pH in *H. pylori* detected? It is important to keep in mind that what induces urease activity may not necessarily be the sensor that detects a decrease in pH. For example, Scenario 4 (Table 2) misses nickel so urease activity would be weak or absent. Adding nickel or inducing the expression of genes involved in nickel acquisition from the environment would induce urease activity. It is partially for this reason that NikR, the nickel-responsive regulator, was hypothesized to be a sensor of pH decrease in *H. pylori* (Carpenter et al., 2015; van Vliet, Ernst, & Kusters, 2004; van Vliet, Kuipers, Stoof, Poppelaars, & Kusters, 2004). However, an accurate sensor of pH change should either sense the pH directly or sense a quantity that is perfectly associated with pH change. Take for example a related pathogen, *H. hepaticus*. It should have urease activity during its transition from stomach to liver. Once it arrives at the liver, there is no further need for urease.

A good sensor that would outcompete alternative mutants should either sense pH directly, or some quantity that can definitely tell the difference between stomach and liver. In this context it makes sense that *H. hepaticus* has a sensor, the transcriptional regulator Fur, for iron concentration (Belzer, van Schendel, Kuipers, Kusters, & van Vliet, 2007). High iron concentration triggers Fur to bind 41-nt iron response element upstream of *ureAB* and inhibits its transcription (Belzer et al., 2007). Because liver tissues contain much more iron than stomach, such a regulator would ensure high urease activity in the stomach but not in liver when it is not needed. However, even this mechanism may not work perfectly in *H. hepaticus*. If the host eats a large chunk of iron-rich pork liver, then *H. hepaticus* would mistake the host stomach for a liver. Similarly, nickel level should not be used as a switch for urease activity because it is not well correlated with pH level. It is therefore not surprising that *H. pylori* does not use nickel availability to regulate urease activity. The presence or absence of Fur in *H. pylori* is associated with changed expression of many genes (Gancz, Censini, & Merrell, 2006), but there is no clear indication that the changed expression pattern is related to pH homeostasis.

Several two-component systems (TCSs) have been documented in *H. pylori* that sense and respond to environmental changes (Beier & Frank, 2000; Scarlato, Delany, Spohn, & Beier, 2001). Three TCSs are directly, and probably synergistically (Carpenter et al., 2015; Loh & Cover, 2006; Scott et al., 2010; Wen et al., 2009; Wen, Feng, Scott, Marcus, & Sachs, 2006, 2007), related to pH homeostasis in *H. pylori*. The arsRS TCS (Loh et al., 2021; Müller, Götz, & Beier, 2009; Pflock et al., 2006; Pflock, Dietz, Schär, & Beier, 2004; Pflock, Kennard, Delany, Scarlato, & Beier, 2005; Pflock, Kennard, Finsterer, & Beier, 2006) is the best studied of the three, but the flgRS is also well substantiated (Scott et al., 2010; Wen et al., 2009).

The third TCS involved in urease regulation is crdRS (Loh & Cover, 2006) which may serve as an alternative regulatory mechanism for pH homeostasis. The well documented function of crdRS is to maintain copper homeostasis (Waidner, Melchers, Stähler, Kist, & Bereswill, 2005). Its relevance to regulating urease activity is not well-established. Full urease activity was induced in crdRS null mutants of *H. pylori* (Pflock, Müller, & Beier, 2007). Therefore, while one cannot exclude the possibility of crdRS as an accessary system for acid resistance, it is clearly not essential for acid resistance.

4.2 Effectors controlling the expression of the urease gene cluster

pH-gated transcription of urease gene cluster is mediated by arsRS and flgRS. ArsR is essential for survival in *H. pylori* (Beier & Frank, *2000;* McDaniel, Dewalt, Salama, & Falkow, 2001). Experimental results suggests that protonation of some residues in ArsS results in changes in ArsS conformation (Müller et al., 2009), which may facilitate its autophosphorylation. This is followed by the transfer of the phosphoryl group to the cognate response regulator ArsR which binds to regulatory motifs upstream of the *ureAB* to initiate downstream gene regulation (Gupta, Borin, Cover, & Krezel, 2009). Comparison of transcriptomic data established ArsR as a regulator of gene expression (Bury-Moné et al., 2001; Dietz, Gerlach, & Beier, 2002; Forsyth, Cao, Garcia, Hall, & Cover, 2002; Merrell, Goodrich, Otto, Tompkins, & Falkow, 2003). A low pH also upregulates *arsS* and *arsR* expression (Wen et al., 2006).

Which histidine residue in ArsS changes its phosphorylation state in response to pH change is not fully established. One of the best studied TCSs consists of the histidine kinase HK853 (gene ID: THMC_RS04370; current locus_tag: THEMA_00370) and the response regulator RR468 (gene ID: TM_RS02395; current locus_tag: THEMA_RS02355) in the *Thermotoga maritima* reference genome (NC_023151) (Casino, Fernández-Alvarez, Alfonso, Rivas, & Marina, 2007). The histidine residue phosphorylated is at site 260 in an H-D/E-L-R-T-P motif highly conserved among histidine kinases in bacterial TCSs (Casino et al., 2007; Ji et al., 2020; Liu et al., 2017; Mideros-Mora, Miguel-Romero, Felipe-Ruiz, Casino, & Marina, 2020; Willett & Kirby, 2012). All *H. pylori* ArsS proteins contain H214 in the same motif **H-E-L**-R-**T-P** where the five bolded amino acids are invariant among 217 ArsS proteins annotated in *H. pylori* RefSeq genomes. Unfortunately, a mutation experiment examined seven histidine residues upstream of H214 but not H214 itself (Müller et al., 2009). The authors suggested that the regulatory site might be somewhere else. However, some experimental tests of pH-sensing share a problem of reasoning. The expression of the regulated genes is typically hypothesized to differ at low pH between the wild type and the mutant. However, the regulator may act when pH is neutral to repress urease activities. At low pH, the regulatory effect is removed so the originally repressed genes are expressed. This would result in no difference in gene expression at low pH, i.e., the regulatory repression is removed in the wild type and disabled in the mutant.

It is important to contrast the expression of regulated genes between neutral and acidic pH instead of just contrasting the expression of regulated genes between the wild type and the mutant at acidic pH.

The aspartate residue in ArsR that receives the phosphate group from the phosphorylated histidine in ArsS is the highly conserved D52 (Loh et al., 2021; Pflock, Finsterer, et al., 2006; Wen et al., 2006). This is also consistent with the well-studied HK853-RR468 TCS (Liu et al., 2017; Mideros-Mora et al., 2020) where D53 in RR468 corresponds to D52 in ArsR (Fig. 2). However, mutating D52 to asparagine has little phenotypic effect (Schär, Sickmann, & Beier, 2005), suggesting that the function of AsrR is independent of its phosphorylation states. However, the study did not contrast gene expression patterns between neutral and acidic pH.

Among genes regulated by ArsR are *ureAB* (encoding urease), *hpn*, *hpnL*, and *hspA* (encoding nickel-binding proteins that are directly related to urease activity), and *uroF* (encoding arginase that generates urea from arginine as a supplemental source of urea) (de Reuse et al., 2013; Loh et al., 2021; Pflock, Finsterer, et al., 2006; Vinella et al., 2015). These target genes are induced by ArsR upon exposure of *H. pylori* to low pH (Loh et al., 2021; Pflock, Finsterer, et al., 2006). Therefore, the arsRS system contributes to both essential components of urease activity, urease gene expression and nickel homeostasis. The arsRS TCS controls the expression of as many as 178 genes related to pH homeostasis (Loh et al., 2021), but its regulatory function is perhaps the most visible on the expression of *ureAB* (Carpenter et al., 2015). The basal level of *ureAB* transcription is reduced when ArsS or ArsR is absent (Carpenter et al., 2015).

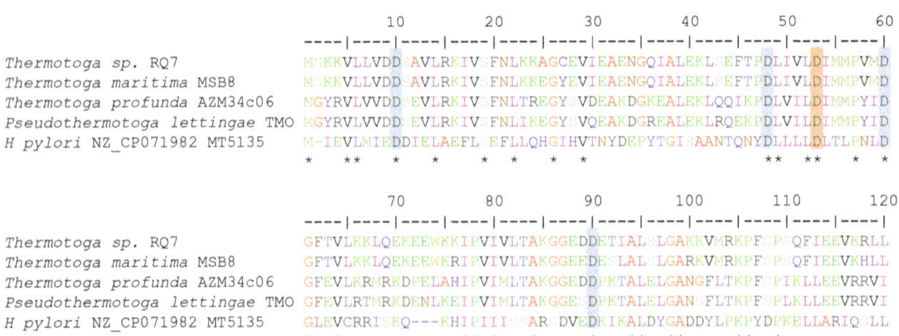

Fig. 2 Partial alignment of ArsR from *H. pylori* strain MT5135 (NZ_CP071982) against RR468 homologs. D53 in RR468 homolog (the first four sequences) receives the phosphate group from the phospshorylated H260 in histidine kinase HK853, and corresponds to D52 in ArsR in *H. pylori* (the last sequence). Other conserved D residues are shaded in blue.

There are two ArsR binding sites (shaded in pink in Fig. 3) immediately upstream of *ureAB* (Fig. 3). The first is conserved among the three *H. pylori* species, and the second is conserved not only in the three *H. pylori* strains, but also in *H. acinonychis* parasitizing big cats (Dailidiene et al., 2004) and *H. cetorum* parasitizing whales (Davison et al., 2014; Harper et al., 2002) (Fig. 3).

DNA palindromes and other secondary structural motifs, formed after the separation of the two strands, are often recognized as cleavage sites or transcription factor binding sites (Ganapathiraju, Subramanian, Chaparala, & Karunakaran, 2020; Subramanian, Chaparala, Avali, & Ganapathiraju, 2016). The second ArsR binding motif (Fig. 3) forms a hairpin in all three representative *H. pylori* strains (Fig. 4). Four nucleotide sites differ among the three sequences and are colored in pink, blue, red, and green, respectively (Fig. 4). However, none of these nucleotide replacements alters the secondary structure. It is likely that the secondary structure represents a

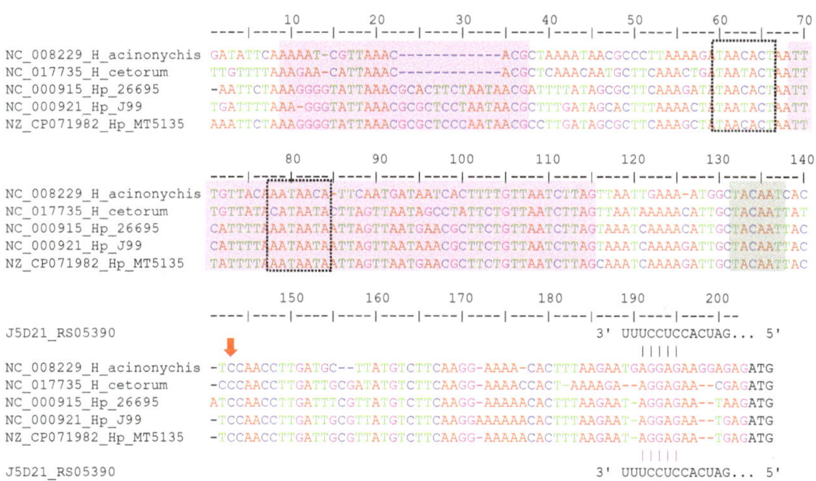

Fig. 3 Upstream sequences of ureA genes from closely related species in *Helicobacter*, showing the recognition sites for ArsR and NikR based on *H. pylori* 26695. Redrawn from Carpenter et al. (2015) that incorporates experimental results from several studies (Dosanjh, Hammerbacher, & Michel, 2007; Evans & Michel, 2012; Pflock et al., 2005). Shaded in pink is the binding sites protected by the binding of ArsR∼P. The boxes with dashed lines designate binding sites for NikR. Shaded in gray is the -10 site that supports the basal level of ureAB. The red arrow points to the transcription initiation site. Start codon AUG is shown at the 3′ end. Immediately upstream is the Shine-Dalgarno sequence pairing with the 3′ tail of the 16S ribosomal RNA (J5D21_RS05390). Sequence names are in the format of GenBank accession_species name. Hp: *Helicobacter pylori*. H: *Helicobactor*.

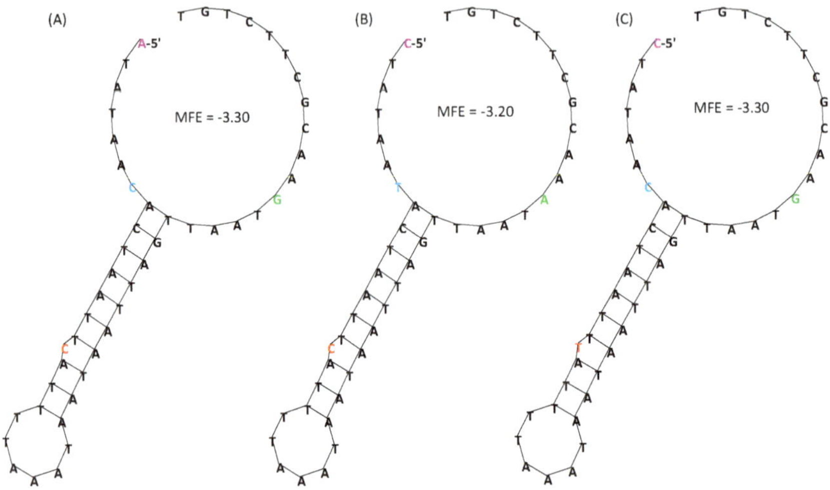

Fig. 4 Hairpin structure that could form at the second ArsR binding motif of the *ureA* gene in the three *H. pylori* strains in Fig. 3, after the DNA melts into separate strands so that each strand can self-anneal. (A) Strain 26695 (which is identical to Strain Puno135 in this sequence segment). (B) Strain J99. (C) Strain MT5135. The three sequences differ in only four homologous sites, highlighted in pink, blue, red and green, respectively. Only one strand of the double-stranded DNA is shown. Generated from DAMBE (Xia, 2018b) which implements the Vienna RNA folding library (Hofacker, 2003). This is an extrapolation from RNA to DNA, with a folding temperature at 37 °C.

recognition signal for ArsR so mutations at the stem region are selected against. To validate the functional importance of these hairpins, one needs to disrupt the base-pairing and to observe if ArsR regulation of *ureAB* transcription is altered. A bioinformatic alternative is to align many homologous sequences from various *H. pylori* strains to see if the stem region is far more conserved than the flanking region, and if a nucleotide substitution disrupting the stem would lead to a compensating nucleotide restoring the base-pairing.

UreA and UreB were once considered constitutively expressed. The gene regulation network involving arsRS, as well as empirical characterization of gene expression (Carpenter et al., 2015; Scott, Marcus, Wen, Oh, & Sachs, 2007), revealed a more dynamic nature of urease gene expression at different pH. There is also a clear pattern of differential degradation of mRNAs from the urease gene cluster, with greater degradation at neutral pH than at low pH (Akada et al., 2000).

Urease maturation requires accessary proteins UreE-F-G-H to insert nickel ions into urease, and urease activity depends on UreI that dramatically increased urea uptake upon a reduction of pH (Scott et al., 2010; Weeks & Sachs, 2001; Wen et al., 2009). While arsRS can sense pH change and regulate the expression of many genes, there is no direct evidence that it regulates the expression of *ureIEFGH* which appears to be on the same operon separate from the *ureAB* operon (Akada et al., 2000). In contrast, flgRS, which was originally thought to function only in motility and flagellar biogenesis, can not only sense pH change, but also regulate the expression of *ureAB* as well as *ureIEFGH* (Scott et al., 2010; Wen et al., 2009).

4.3 Nickel homeostasis and urease activation

There are two essential enzymes in *H. pylori* that require nickel to function: urease and hydrogenase (de Reuse et al., 2013; Schauer et al., 2010; Vinella et al., 2015). Therefore, it is important to maintain cytoplasmic nickel availability (Stingl & De Reuse, 2005). Nickel is imported into periplasm by an acid-gated transporter NixA and stored in H-rich proteins such as Hpn and HspA. Cytoplasmic nickel homeostasis is maintained by NikR, the nickel-responsive regulator belonging to the ribbon-helix-helix family of transcription factors (Ernst et al., 2005). NikR controls the expression of nickel importing and storage proteins (Bury-Moné et al., 2004; Ernst et al., 2005; Stingl & De Reuse, 2005; Suerbaum, Thiberge, Kansau, Ferrero, & Labigne, 1994; van Vliet et al., 2001, 2002).

It is important to maintain sufficient amount of nickel in *H. pylori* cells. Daughter cells with insufficient nickel for urease activation would not be able to survive in acidic environment (Stingl & De Reuse, 2005). The intracellular nickel concentration in *H. pylori* is about 50 times higher than that in *E. coli* (de Reuse et al., 2013). Therefore, a *H. pylori* cell needs enough nickel not only to maintain pH homeostasis to survive, but also to ensure the survival of its daughter cells (Gancz et al., 2006).

Different nickel-sequestrating proteins contribute to a common pool of nickel used by urease and hydrogenase to function. Why cannot there be separate pools of nickel with one pool for urease and another for hydrogenase? For example, nickel-storage proteins Hpn and HpnL/Hpn-2 may supply nickel to urease accessary proteins UreE, UreF, UreG and UreD/UreH which in turn would insert nickel into urease. In contrast, nickel-storage proteins HspA and HspB may supply nickel to hydrogenase accessary proteins HypA and HypB which in turn would insert nickel into hydrogenase. The reason against such separate pools is that most *H. pylori* strains do not have Hpn proteins. For example, the reference genome

NZ_CP071982 from strain MT5135 does not have an *hpn* gene. Only 20 of 338 *H. pylori* RefSeq genomes have an *hpn* gene. However, all 338 *H. pylori* RefSeq genomes have *hspA* and *hspB* genes, suggesting that HspA and HspB represent a more ancient and universal mechanism of nickel storage and release in *H. pylori* than Hpn and HpnL/Hpn-2. I will discuss this in more detail in following sections.

4.3.1 Nickel importing and exporting

The membrane protein NixA (encoded by *nixA*, locus_tag J5D21_RS01795 in the reference genome) imports nickel into *H. pylori* cells (Fulkerson & Mobley, 2000), although there might be alternative nickel transporters as reviewed previously (de Reuse et al., 2013). Surplus nickel not sequestered by nickel storage proteins will bind to NikR (Stingl & De Reuse, 2005). The binding of Ni^{+2} may facilitate the electrostatic interaction between NikR and the negatively charged DNA backbone. Ni^{+2}NikR binds to the promoter of *nixA* to repress its transcription (de Reuse et al., 2013; Ernst et al., 2005; Wolfram, Haas, & Bauerfeind, 2006), thereby reducing nickel import. Ni^{+2}NikR also binds to the promoter of *hspA* and *hpn* to increase the production of nickel storage proteins HspA and Hpn (Muller et al., 2011), thereby increasing the nickel storage capacity. The decrease in nickel import and the increase in nickel storage avoid cytotoxicity of surplus nickel or accidental activation of urease activity under neutral pH.

Another mechanism for preventing surplus nickel is a nickel export mechanism through an efflux pump (Stähler et al., 2006) involving three genes in the *H. pylori* strain 26695. These three genes have locus_tag HP0969, HP0970 and HP0971 in the now obsolete GenBank sequence NC_000915. They correspond to genes with locus_tag HP1CH_RS04960, HP1CH_RS04965, and HP1CH_RS04970, respectively, in the new GenBank genome record for strain 26696, NC_018939. Free nickel ions are prone to trigger urease activities which should be avoided under neutral pH. The efflux pump exports surplus nickel that cannot be sequestered by nickel storage proteins.

4.3.2 Nickel sequestrating and storage proteins

There are two important evolutionary events in *H. pylori* related to nickel storage. The first is an acquisition of a C-terminus extension (Fig. 5) of the GroES protein. This C-terminus extension converts GroES to a nickel storage protein (Schauer et al., 2010; Suerbaum et al., 1994). Although all

```
                                 10        20        30        40        50        60
                        ----|----|----|----|----|----|----|----|----|----|----|----|
NC_008229_H_acinonychis MKFQPLGERVLVERLEEENKTSSGIIIPDNAKEKPLMGVVKAVSHKISEGCKCVKEGDVI
NC_017735_H_cetorum     MKFQPLGERVLVERLEEENKTSSGIIIPDNAKEKPLMGVVKAVSHKISEGCKCVKEGDVI
NC_017379_Hp_Puno135    MKFQPLGERVLVERLEEENKTSSGIIIPDNAKEKPLMGVVKAVSHKISEGCKCVKEGDVI
Hp_26695_J99_MT5135(1)  MKFQPLGERVLVERLEEENKTSSGIIIPDNAKEKPLMGVVKAVSHKISEGCKCVKEGDVI
NC_004917_H_hepaticus   MKFRPLGERVLVERVEEDTKTSSGIIIPDNAKEKPLMGIVKAVSAKIKDD-KILKENDKV
NC_002163_C_jejuni      MNFQPLGKRVLVKRVEETKTTASGIIIPDNAKEKPLMGEVVAVSKEITD----IANGDKI
                        * * *** **** * **   * ************* ** * ***  *          *

                                 70        80        90       100       110
                        ----|----|----|----|----|----|----|----|----|----|----|---
NC_008229_H_acinonychis AFGKYKGTEIVLDGTEYMVLELEDILGIVNAGSCCHANSHDHKDHKHAEACCHDHKKH
NC_017735_H_cetorum     AFGKYKGTEIVLDGTDYMVLELEDILGIVGAGSCCHGGHHDHKHGAEHNSCCHGH---
NC_017379_Hp_Puno135    AFGKYKGAEIVLDGTEYMVLELEDILGIVSSGSCCHTNSHDHKHAKEHEACCHDHKKH
Hp_26695_J99_MT5135(1)  AFGKYKGAEIVLDGTEYMVLELEDILGIVGSGSCCHTGNHDHKHAKEHEACCHDHKKH
NC_004917_H_hepaticus   VFGKYKGAEIKLDSKEFIVLELDDILGVIEK---------------------------
NC_002163_C_jejuni      VFAKYGGTEIKLDNNEYLVLNLDDILGILK----------------------------
                         * ** * ** **     ** * * **
```

Fig. 5 Gastric pathogens (the first four) share a C-terminus extension in GroES/HspA that is missing in non-gastric pathogens (the last two, *Helicobacter hepaticus* and *Campylobacter jejuni*). Hp_26695_J99_MT5135 represents *H. pylori* strains 26695, J99 and MT5135. The extension has two CC residues (shaded in blue) and is rich in histidine.

seven proteins in Fig. 5 are annotated in GenBank files as encoded by the *groES* gene, *H. pylori* literature names this gene *hspA* encoding the HspA protein with two domains, the GroES-like domain A shared among all seven proteins and the nickel-binding domain B which is the C-terminus extension shared among the first four gastric pathogens (Fig. 5). All 338 *H. pylori* RefSeq genomes have *hspA* and *hspB* genes. All *hspA* genes feature the conserved groES-like domain A and the nickel-binding domain B (Fig. 5), and they all share perfectly conserved CC doublets (Fig. 5). The annotated *groL* gene is referred to in *H. pylori* literature as *hspB*.

Two features are prominent in this C-terminus extension (domain B) of HspA. The first is histidine-rich, and the second is the two CC residues (shaded in blue, Fig. 5). Histidine residues are crucial for Ni^{2+} binding (Bellucci, Zambelli, Musiani, Turano, & Ciurli, 2009). In fact, the HX_4DH motif immediately following the first CC in the domain B of HspA (Fig. 5) is sometimes considered as the signature of nickel binding protein in the four *H. pylori* strains. Point mutations either at the last DH of the motif or the CC before the motif, or deletion of domain B, would impair nickel storage by HspA (Schauer et al., 2010). Such mutants also became more prone to nickel cytotoxicity (Schauer et al., 2010).

The second evolutionary event in nickel storage is a gene duplication during the early evolution of *H. pylori* that gave rise to paralogues *hpn* and *hpnL* (Fig. 6) with strong effect on nickel insertion and urease maturation (de Reuse et al., 2013; Ge et al., 2006; Gilbert, Ramakrishna,

```
                                    10        20        30        40
                            ----|----|----|----|----|----|----|----|
NC_000915_Hp_26695_hpnL     MAHHE------QQQQQQANSQHHHHHHAHHHHYYGGEHHH
NC_000921_Hp_J99_hpnL       MAHHEQQHQA-QQQQQQANSQHHHHHHVHHHHYYGGEHHH
NC_017379_Hp_Puno135_hpnL   MAH----------QQQQANSQHHHHHHAHHHHYYGGEHHH
NZ_CP071982_Hp_MT5135_hpnL  MAHHEQQHQAQQQQQQQANSQHHHHHHAHHHHYYGGEHHH
NC_000915_Hp_26695_hpn      MAHHEEQH----------GGHHHHHHHTHHHHYHGGEHHH
NC_000921_Hp_J99_hpn        MAHHEEQH----------GGHHHHHHHTHHHHYHGGEHHH
NC_017379_Hp_Puno135_hpn    MAHHEEQH----------GGHHHHHHHTHHHHYHGGEHHH
                            ***                  ****** ***** ******

                                    50        60        70
                            ----|----|----|----|----|----|----|---
NC_000915_Hp_26695_hpnL     HNAQQHAEQQAEQQAQQQQQQQAHQQQQQKAQQQNQQY
NC_000921_Hp_J99_hpnL       HNAQQHAEQQAEQHAQQQQQQQAHQQQQQKAQQQNQQY
NC_017379_Hp_Puno135_hpnL   HNAQQHAEQQAEQQVQQQAQ----QQQQQQAQQQNQQY
NZ_CP071982_Hp_MT5135_hpnL  HNAQQHAEQQAEQHAQ--------QQQQQKAQQQNQQY
NC_000915_Hp_26695_hpn      HHHSSHHEEGCCSTSD--------SHHQEEGCCHGHHE
NC_000921_Hp_J99_hpn        HHHSSHHEEGCCSTSD--------SHHQEEGCCHGHHE
NC_017379_Hp_Puno135_hpn    HHHSSHHEEGCCSTSD--------SHHQEEGCCHGHHE
                            *    * *                   *
```

Fig. 6 Protein sequence alignment of paralogous genes *hpnL* (first four) and *hpn* (last three). The two CC residues are shaded in blue.

Sunderman, Wright, & Plaut, 1995; Rowinska-Zyrek, Witkowska, Bielinska, Kamysz, & Kozlowski, 2011; Vinella et al., 2015; Zeng, Yang, & Sun, 2011; Zeng, Zhang, Li, & Sun, 2008). In strain J99, the *hpn* gene is immediately downstream of *hpnL* (most likely in the same operon), but the two genes are separated by about 5000 bases in strains 26695 and Puno135. Among the 338 *H. pylori* RefSeq genomes, 219 have a homologue of *hpnL*, but only 20 have a homologue of *hpn*. All genomes with an *hpn* also have an *hpnL*. One may infer that *hpnL* represents an ancestral state and *hpn* a derived state resulting from neofunctionalization after gene duplication. All *hpn* sequences are more closely related to each other than they are to *hpnL* sequences, suggesting a single gene duplication event instead of independent duplications in different *H. pylori* strains. The *hpnL* gene in the reference genome is often referred to as *hpn-2* in publications (de Reuse et al., 2013; Vinella et al., 2015).

The middle part of the two paralogous genes *hpnL* and *hpn* are conserved in amino acid sequences (Fig. 6), but the flanking sequences have diverged much between the two. The C-terminus of Hpn share two features with the domain B of HspA (Fig. 5) in that they are both rich in histidine residues and features the two CC residues. These two features are considered signatures of nickel-binding proteins (Schauer et al., 2010).

Both *hpn* and *hspA* mutants of *H. pylori* are sensitive to exogeneous nickel concentration than their wild type (Mobley et al., 1999; Seshadri, Benoit, & Maier, 2007). This is consistent with the interpretation that they are the major nickel storage proteins in *H. pylori*. HspA and Hpn bind directly to nickel but their respective partners, HspB and HpnL/Hpn-2, do not. However, HspB and HpnL/Hpn-2 may facilitate the transfer of nickel to nickel-inserting accessory proteins (UreEFGH for urease and HypAB for hydrogenase.

Both nickel storage proteins (Hpn and HspA) are rich in histidine residues which are inherent sensors for pH changes. Compiled pKa values for histidine residue range from 6.0 to 6.5 (Xia, 2018a). Thus, when pH changes from 7 to below 6, histidine residues will change state from unprotonated to protonated. In contrast, lysine and arginine residues have pK_a values of 10.0 and 12.5, respectively, i.e., they are protonated under all physiological conditions. This feature of histidine residues as a pH sensor is reflected in the protein isoelectric point, being 6.8 for Hpn and 6.6 for HspA. This means that they tend to be negatively charged at $pH > 7$ and would electrostatically interact favorably with Ni^{2+}. When *H. pylori* comes into contact with gastric fluid with $pH < 6$, the two proteins will carry net positive charges, and would tend to release Ni^{2+} because of electric repulsion. The free Ni^{2+} can then be used to activate urease. This hypothesis of pH-mediated sequestration and release of Ni^{2+} by the nickel binding proteins represents another mechanism of maintaining pH homeostasis and should be tested experimentally.

4.3.3 Different pathways for nickel insertion between urease and hydrogenase

The stored nickel in HspA and Hpn will be inserted into urease and hydrogenase, assisted by their respective partner HspB and Hpn-2. Nickel insertion requires a metallochaperone and a GTPase (Biagi, Musiani, & Ciurli, 2013). HypA and HypB are the metallochaperone and GTPase, respectively, for inserting nickel into hydrogenase (Benoit, McMurry, Hill, & Maier, 2012; Fernández-Bravo, López-Fernández, & Figueras, 2019; Xia, Li, Sze, & Sun, 2009; Xia, Li, Yang, Wong, & Sun, 2012). UreE and UreG are the metallochaperone and the GTPase, respectively, for inserting nickel into urease, with UreD and UreF physically interact with UreG to facilitate GTP hydrolysis by UreG (Biagi et al., 2013; Yang, Li, Lai, & Sun, 2015). However, UreE dimer and HypA can form a hetero-complex

```
                                   60        70        80        90        100       110
                             ----|----|----|----|----|----|----|----|----|----|----|----|----|--
NC_008229_H. acinonychis     FREESLVCKDAVLDIVDEKVELECKDCSHVFKPNALDYGVCEKCHSKNVVITQGNEMRLLSLEMLAE
NC_017735_H. cetorum         FREESLVCKDAILDIVDEKIELECKDCSHVFKPNTLDYGVCEQCQGKNISVIQGNEMRLLSLEMLAE
NC_000915_Hp_26695_J99       FREESLVCKDAILDIVDEKVELECKDCSHVFKPNALDYGVCEKCHSKNVIITQGNEMRLLSLEMLAE
NC_017379_Hp_Puno135         FREESLVCKDAILDIVDEKVELECKDCSHVFKPNALDYGVCEKCHSKNVVITQGNEMRLLSLEMLAE
NZ_CP071982_Hp_MT5135        FREESLVCKDAILDIVDEKVELECKDCSHVFKPNALDYGVCEKCHSKNVIITQGNEMRLLSLEMLAE
                             *********** ******* ************** ******* *  **    ***************
```

Fig. 7 Two metal-binding sites in HypA shaded in red. The first 50 sites were not shown. Both sites feature the conserved CXXC with a downstream histidine residue. Hp: *Helicobacter pylori*. Strains 26695 and J99 have identical HypA sequences.

HypA-$(UreE)_2$ and may serve as co-metallochaperones (Hu, Huang, & Maroney, 2018; Yang et al., 2014) in which nickel is transferred from HypA to UreE (Yang et al., 2014).

There are two metal-binding sites in HypA that can also bind to Zn^{+2} (Fig. 7). Both feature the CXXC motif with a downstream histidine residue (Fig. 7) that are crucial for metal-binding (Kolkowska et al., 2015; Spronk et al., 2018). The N-terminal site is more conserved (Fig. 7), and binds to Zn^{+2} and Ni^{+2} more effectively (Kolkowska et al., 2015), than the C-terminal site. The P83 site, as well as the distance between the two sites, can affect metal-binding ability of HypA (Kolkowska et al., 2015).

Although all details have not been worked out, the working hypothesis is that HypA + HypB receives Ni^{2+} from HspA + HspB, but UreE + UreG can get Ni^{2+} from either HspA + HspB, Hpn + HpnL, or HypA. This hypothesis predicts that disabling the function of HspA by mutation would have a strong effect on the HspA + HspB nickel uptake and hydrogenase activity but only a moderate effect on the UreE + UreG nickel uptake and urease activity if Hpn + HpnL is present. This prediction is consistent with the mutation experiments with HspA (Schauer et al., 2010) because UreE + UreG can obtain nickel from an alternative source, i.e., from Hpn + HpnL. Similarly, *Δhpn* or *Δhpn* + *ΔhpnL* mutant is expected to only affect urease activity slightly because UreE + UreG can obtain nickel from HspA + HspB, which is also consistent with the experimental results (Gilbert et al., 1995). This suggests that the function of HypA and Hpn are redundant from the urease perspective. Thus, the two nickel insertion pathways are intertwined and that HypA and HypB are also involved in urease activation (Benanti & Chivers, 2009; de Reuse et al., 2013; Herbst et al., 2010; Olson, Mehta, & Maier, 2001; Vinella et al., 2015). There has been no study that knocks out both *hypA* and *hpn* genes. One would expect a dramatic reduction in urease activity with such a double knockout unless there are unknown proteins serving the same redundant function.

Nickel insertion into the UreA subunits in the mature urease is mediated by the UreEFGH complex (Bellucci et al., 2009; de Reuse et al., 2013; Stingl et al., 2008). The hypothesized process of nickel insertion into urease may be described in broad strokes as follows. The metallochaperone UreE receives Ni^{2+} from the Ni^{2+} pool represented by HspA + HspB and Hpn + HpnL nickel storage proteins, and passes the nickel to UreG, leading to UreG dimerization. The UreG dimer with Ni^{2+} then bind to UreFH which helps to pry open the large urease and allows Ni^{2+} from UreG dimer to access the space between UreAB subunits. UreG then dissociates from UreFH to repeat the process.

Electrostatic interaction between UreA and UreB may also play a role in urease maturation via nickel insertion. The isoelectric point (pI) is 9.00 for UreA and 5.93 for UreB. At neutral pH, UreA is positively charged and UreB negatively charged, the two being therefore attracted to each other electrostatically. In the acidic gastric fluid in which the cellular cytoplasm is 5 or lower, both UreA and UreB would have net positive charge and will not bind tightly to each other. This would facilitate the opening of the inner space for Ni^{2+} insertion.

4.3.4 Regulatory proteins for maintaining nickel homeostasis

Urease is a large enzyme in the form of $((UreAB)_3)_4$. Each UreAB dimer in activated form harbors two nickel ions, so that the activated urease contains 24 nickel ions (Ha et al., 2001). Nickel imported into the cell will bind to nickel storage proteins Hpn and HspA as reviewed in the previous section. Surplus nickel that cannot be sequestered by the nickel storage proteins will bind to NikR which will bind to the promoter of *hpn*, *hpnL* and *hspA* to increase their expression to enhance the nickel storage capacity (Muller et al., 2011). Ni^{+2}NikR will also bind to the promotor of *nixA* to suppress its transcription, thereby reducing the import of nickel into the cell. However, NikR is a pleiotropic transcription factor affecting the expression of many other genes (Contreras, Thiberge, Mandrand-Berthelot, & Labigne, 2003; Dosanjh, West, & Michel, 2009; Loh et al., 2021). NikR also has two binding motifs upstream of the *ureAB* gene (Fig. 3). In particular, apo-NikR does not bind to the promoter of *ureAB*, but Ni^{+2}NikR does (Dosanjh et al., 2007) and increases the expression of *ureAB* transcription (Carpenter et al., 2015; Jones, Li, & Zamble, 2018; Loh et al., 2021). The dual function of both arsRS and NikR in regulating urease gene expression and nickel homeostasis indicate crosstalk between the two (Carpenter et al., 2015; van Vliet, Ernst, & Kusters, 2004; van Vliet, Kuipers, et al., 2004).

The function of NikR differs fundamentally between *H. pylori* and *H. hepaticus*. NikR in the latter appears to have the sole function of inhibiting nickel-dependent urease maturation (Benoit, Seshadri, Lamichhane-Khadka, & Maier, 2013). Disabling NikR leads to dramatic increase in urease activity without changing total urease level. *H. hepaticus* is a liver pathogen with no need for acid-resistance. Urease activity at neutral pH would lead to overalkalization and ammonia toxicity and should therefore be suppressed.

While many gene regulation pathways might be interpreted to serve certain functions, one should not assume that all gene regulations involving arsRS or NikR should be beneficial to *H. pylori*. All gene regulation networks in nature are not designed but evolved. When a transcription factor has evolved a new binding site and gained a new set of regulated genes, this altered gene regulation in most cases are expected to be deleterious and eliminated by purifying selection. In the rare cases where the altered gene regulation is overall beneficial and favored by natural selection, not all regulated genes will contribute positively to the organism's fitness. Even a designed drug often has unintended or unavoidable side effects. Suppose the following observations: (1) a transcription factor X has function F and (2) N genes are regulated by X. A researcher would often make far-fetched arguments that all N genes are contributing, directly or indirectly, to accomplish function F. Many of these N genes may just happen to harbor a motif near their promotor for X to bind to. Their regulation may have no functional implication or even slightly deleterious. An altered gene regulation pattern would be favored as long as the benefit overweighs the cost.

Acknowledgments

This research was funded by Discovery Grant from Natural Science and Engineering Research Council (NSERC) of Canada, grant number RGPIN/2018-03878. I thank Jeremy Xia for helping with secondary structure, data compilation and comments on an early version of the manuscript.

References

Akada, J. K., Ogura, K., Dailidiene, D., Dailide, G., Cheverud, J. M., & Berg, D. E. (2003). *Microbiology*, *149*, 1901–1909.

Akada, J. K., Shirai, M., Takeuchi, H., Tsuda, M., & Nakazawa, T. (2000). *Molecular Microbiology*, *36*, 1071–1084.

Allen, A., Flemstrom, G., Garner, A., & Kivilaakso, E. (1993). *Physiological Reviews*, *73*, 823–857.

Alm, R. A., Bina, J., Andrews, B. M., Doig, P., Hancock, R. E., & Trust, T. J. (2000). *Infection and Immunity*, *68*, 4155–4168.
Andersen, A. P., Elliott, D. A., Lawson, M., Barland, P., Hatcher, V. B., & Puszkin, E. G. (1997). *Journal of Clinical Microbiology*, *35*, 2918–2922.
Baik, S. C., Kim, K. M., Song, S. M., Kim, D. S., Jun, J. S., Lee, S. G., et al. (2004). *Journal of Bacteriology*, *186*, 949–955.
Bauerfeind, P., Garner, R., Dunn, B. E., & Mobley, H. L. (1997). *Gut*, *40*, 25–30.
Baumgartner, H. K., & Montrose, M. H. (2004). *Gastroenterology*, *126*, 774–783.
Beasley, D. E., Koltz, A. M., Lambert, J. E., Fierer, N., & Dunn, R. R. (2015). *PLoS One*, *10*, e0134116.
Beier, D., & Frank, R. (2000). *Journal of Bacteriology*, *182*, 2068–2076.
Bellucci, M., Zambelli, B., Musiani, F., Turano, P., & Ciurli, S. (2009). *The Biochemical Journal*, *422*, 91–100.
Belzer, C., van Schendel, B. A., Kuipers, E. J., Kusters, J. G., & van Vliet, A. H. (2007). *Infection and Immunity*, *75*, 745–752.
Benanti, E. L., & Chivers, P. T. (2009). *Journal of Bacteriology*, *191*, 2405–2408.
Benoit, S. L., McMurry, J. L., Hill, S. A., & Maier, R. J. (2012). *Biochimica et Biophysica Acta*, *1820*, 1519–1525.
Benoit, S. L., Seshadri, S., Lamichhane-Khadka, R., & Maier, R. J. (2013). *Microbiology (Reading, England)*, *159*, 136–146.
Biagi, F., Musiani, F., & Ciurli, S. (2013). *Journal of Biological Inorganic Chemistry*, *18*, 571–577.
Bury-Moné, S., Skouloubris, S., Labigne, A., & De Reuse, H. (2001). *Molecular Microbiology*, *42*, 1021–1034.
Bury-Moné, S., Thiberge, J. M., Contreras, M., Maitournam, A., Labigne, A., & De Reuse, H. (2004). *Molecular Microbiology*, *53*, 623–638.
Capra, E. J., & Laub, M. T. (2012). *Annual Review of Microbiology*, *66*, 325–347.
Carpenter, B. M., West, A. L., Gancz, H., Servetas, S. L., Pich, O. Q., Gilbreath, J. J., et al. (2015). *Frontiers in Microbiology*, *6*, 558.
Carter, E. L., Flugga, N., Boer, J. L., Mulrooney, S. B., & Hausinger, R. P. (2009). *Metallomics: Integrated Biometal Science*, *1*, 207–221.
Casino, P., Fernández-Alvarez, A., Alfonso, C., Rivas, G., & Marina, A. (2007). *Biochimica et Biophysica Acta*, *1774*, 603–609.
Clyne, M., Labigne, A., & Drumm, B. (1995). *Infection and Immunity*, *63*, 1669–1673.
Contreras, M., Thiberge, J. M., Mandrand-Berthelot, M. A., & Labigne, A. (2003). *Molecular Microbiology*, *49*, 947–963.
Correa, P. (1997). *Journal of Physiology and Pharmacology*, *48*, 19–24.
Cussac, V., Ferrero, R. L., & Labigne, A. (1992). *Journal of Bacteriology*, *174*, 2466–2473.
Dailidiene, D., Dailide, G., Ogura, K., Zhang, M., Mukhopadhyay, A. K., Eaton, K. A., et al. (2004). *Journal of Bacteriology*, *186*, 356–365.
Davies, B. J., de Vries, N., Rijpkema, S. G., van Vliet, A. H. M., Penn, C. W., & Microbiol, F. E. M. S. (2002). *Letters*, *213*, 27–32.
Davison, N. J., Barnett, J. E., Koylass, M., Whatmore, A. M., Perkins, M. W., Deaville, R. C., et al. (2014). *Journal of Wildlife Diseases*, *50*, 431–437.
de Reuse, H., Vinella, D., & Cavazza, C. (2013). *Frontiers in Cellular and Infection Microbiology*, *3*, 94.
Debowski, A. W., Walton, S. M., Chua, E.-G., Tay, A. C.-Y., Liao, T., Lamichhane, B., et al. (2017). *PLoS Pathogens*, *13*, e1006464.
Dietz, P., Gerlach, G., & Beier, D. (2002). *Journal of Bacteriology*, *184*, 350–362.
Dosanjh, N. S., Hammerbacher, N. A., & Michel, S. L. (2007). *Biochemistry (Mosc)*, *46*, 2520–2529.
Dosanjh, N. S., West, A. L., & Michel, S. L. (2009). *Biochemistry (Mosc)*, *48*, 527–536.
Dressman, J. B., Berardi, R. R., Dermentzoglou, L. C., Russell, T. L., Schmaltz, S. P., Barnett, J. L., et al. (1990). *Pharmaceutical Research*, *7*, 756–761.

Eaton, K. A., Brooks, C. L., Morgan, D. R., & Krakowka, S. (1991). *Infection and Immunity, 59*, 2470–2475.

Eaton, K. A., & Krakowka, S. (1994). *Infection and Immunity, 62*, 3604–3607.

Engel, E., Peskoff, A., Kauffman, G. L., Jr., & Grossman, M. I. (1984). *The American Journal of Physiology, 247*, G321–G338.

Ernst, F. D., Kuipers, E. J., Heijens, A., Sarwari, R., Stoof, J., Penn, C. W., et al. (2005). *Infection and Immunity, 73*, 7252–7258.

Evans, S. E., & Michel, S. L. (2012). *Dalton Transactions (Cambridge, England: 2003), 41*, 7946–7951.

Fernández-Bravo, A., López-Fernández, L., & Figueras, M. J. (2019). *Microorganisms*, 7.

Forsyth, M. H., Cao, P., Garcia, P. P., Hall, J. D., & Cover, T. L. (2002). *Journal of Bacteriology, 184*, 4630–4635.

Fulkerson, J. F., Jr., & Mobley, H. L. (2000). *Journal of Bacteriology, 182*, 1722–1730.

Ganapathiraju, M. K., Subramanian, S., Chaparala, S., & Karunakaran, K. B. (2020). *Human Genome Variation*, 7, 40.

Gancz, H., Censini, S., & Merrell, D. S. (2006). *Infection and Immunity, 74*, 602–614.

Ge, R., Watt, R. M., Sun, X., Tanner, J. A., He, Q. Y., Huang, J. D., et al. (2006). *The Biochemical Journal, 393*, 285–293.

Gilbert, J. V., Ramakrishna, J., Sunderman, F. W., Jr., Wright, A., & Plaut, A. G. (1995). *Infection and Immunity, 63*, 2682–2688.

Guindon, S., & Gascuel, O. (2003). *Systematic Biology, 52*, 696–704.

Gupta, S. S., Borin, B. N., Cover, T. L., & Krezel, A. M. (2009). *The Journal of Biological Chemistry, 284*, 6536–6545.

Ha, N. C., Oh, S. T., Sung, J. Y., Cha, K. A., Lee, M. H., & Oh, B. H. (2001). *Nature Structural Biology, 8*, 505–509.

Hamajima, N., Goto, Y., Nishio, K., Tanaka, D., Kawai, S., Sakakibara, H., et al. (2004). *Asian Pacific Journal of Cancer Prevention, 5*, 246–252.

Harper, C. G., Feng, Y., Xu, S., Taylor, N. S., Kinsel, M., Dewhirst, F. E., et al. (2002). *Journal of Clinical Microbiology, 40*, 4536–4543.

Herbst, R. W., Perovic, I., Martin-Diaconescu, V., O'Brien, K., Chivers, P. T., Pochapsky, S. S., et al. (2010). *Journal of the American Chemical Society, 132*, 10338–10351.

Hessey, S. J., Spencer, J., Wyatt, J. I., Sobala, G., Rathbone, B. J., Axon, A. T., et al. (1990). *Gut, 31*, 134–138.

Hofacker, I. L. (2003). *Nucleic Acids Research, 31*, 3429–3431.

Hovey, J. G., Watson, E. L., Langford, M. L., Hildebrandt, E., Bathala, S., Bolland, J. R., et al. (2007). *BMC Microbiology*, 7, 26.

Hu, H. Q., Huang, H. T., & Maroney, M. J. (2018). *Biochemistry (Mosc), 57*, 2932–2942.

Hunt, R. H. (2004). *The American Journal of Medicine, 117*, 86S–91S.

Huynen, M., Dandekar, T., & Bork, P. (1998). *FEBS Letters, 426*, 1–5.

Ji, S., Luo, L., Li, C., Liu, M., Liu, Y., & Jiang, L. (2020). *Biochemical and Biophysical Research Communications, 523*, 733–738.

Jones, M. D., Li, Y., & Zamble, D. B. (2018). *Proceedings of the National Academy of Sciences of the United States of America, 115*, 8966–8971.

Kararli, T. T. (1995). *Biopharmaceutics & Drug Disposition, 16*, 351–380.

Katoh, K., Kuma, K., Miyata, T., & Toh, H. (2005). *Genome Informatics, 16*, 22–33.

Kolkowska, P., Krzywoszynska, K., Potocki, S., Chetana, P. R., Spodzieja, M., Rodziewicz-Motowidlo, S., et al. (2015). *Dalton Transactions (Cambridge, England: 2003), 44*, 9887–9900.

Lewis, R. J., Brannigan, J. A., Muchová, K., Barák, I., & Wilkinson, A. J. (1999). *Journal of Molecular Biology, 294*, 9–15.

Liu, Y., Rose, J., Huang, S., Hu, Y., Wu, Q., Wang, D., et al. (2017). *Nature Communications, 8*, 2104.

Loh, J. T., & Cover, T. L. (2006). *Infection and Immunity, 74*, 3052–3059.
Loh, J. T., Shum, M. V., Jossart, S. D. R., Campbell, A. M., Sawhney, N., McDonald, W. H., et al. (2021). *Infection and Immunity, 89*.
Matin, A., Zychlinsky, E., Keyhan, M., & Sachs, G. (1996). *Infection and Immunity, 64*, 1434–1436.
McDaniel, T. K., Dewalt, K. C., Salama, N. R., & Falkow, S. (2001). *Helicobacter, 6*, 15–23.
McGee, D. J., Zabaleta, J., Viator, R. J., Testerman, T. L., Ochoa, A. C., & Mendz, G. L. (2004). *European Journal of Biochemistry, 271*, 1952–1962.
Menaker, R. J., Sharaf, A. A., & Jones, N. L. (2004). *Current Gastroenterology Reports, 6*, 429–435.
Mendz, G. L., & Hazell, S. L. (1996). *Microbiology, 142*, 2959–2967.
Merrell, D. S., Goodrich, M. L., Otto, G., Tompkins, L. S., & Falkow, S. (2003). *Infection and Immunity, 71*, 3529–3539.
Meyer-Rosberg, K., Scott, D. R., Rex, D., Melchers, K., & Sachs, G. (1996). *Gastroenterology, 111*, 886–900.
Mideros-Mora, C., Miguel-Romero, L., Felipe-Ruiz, A., Casino, P., & Marina, A. (2020). *Nature Communications, 11*, 769.
Miguel-Romero, L., Mideros-Mora, C., Marina, A., & Casino, P. (2020). *Methods in Molecular Biology, 2077*, 121–140.
Mine, T., Muraoka, H., Saika, T., & Kobayashi, I. (2005). *Helicobacter, 10*, 125–131.
Mobley, H. L., Garner, R. M., Chippendale, G. R., Gilbert, J. V., Kane, A. V., & Plaut, A. G. (1999). *Helicobacter, 4*, 162–169.
Mobley, H. L., Hu, L. T., & Foxal, P. A. (1991). *Scandinavian Journal of Gastroenterology. Supplement, 187*, 39–46.
Mobley, H. L., Island, M. D., & Hausinger, R. P. (1995). *Microbiological Reviews, 59*, 451–480.
Morgan, D. R., Freedman, R., Depew, C. E., & Kraft, W. G. (1987). *Journal of Clinical Microbiology, 25*, 2123–2125.
Muller, C., Bahlawane, C., Aubert, S., Delay, C. M., Schauer, K., Michaud-Soret, I., et al. (2011). *Nucleic Acids Research, 39*, 7564–7575.
Müller, S., Götz, M., & Beier, D. (2009). *PLoS One, 4*, e6930.
Olson, J. W., Mehta, N. S., & Maier, R. J. (2001). *Molecular Microbiology, 39*, 176–182.
Ovesen, L., Bendtsen, F., Tage-Jensen, U., Pedersen, N. T., Gram, B. R., & Rune, S. J. (1986). *Gastroenterology, 90*, 958–962.
Pflock, M., Dietz, P., Schär, J., & Beier, D. (2004). *FEMS Microbiology Letters, 234*, 51–61.
Pflock, M., Finsterer, N., Joseph, B., Mollenkopf, H., Meyer, T. F., & Beier, D. (2006). *Journal of Bacteriology, 188*, 3449–3462.
Pflock, M., Kennard, S., Delany, I., Scarlato, V., & Beier, D. (2005). *Infection and Immunity, 73*, 6437–6445.
Pflock, M., Kennard, S., Finsterer, N., & Beier, D. (2006). *Journal of Biotechnology, 126*, 52–60.
Pflock, M., Müller, S., & Beier, D. (2007). *Current Microbiology, 54*, 320–324.
Rektorschek, M., Buhmann, A., Weeks, D., Schwan, D., Bensch, K. W., Eskandari, S., et al. (2000). *Molecular Microbiology, 36*, 141–152.
Rowinska-Zyrek, M., Witkowska, D., Bielinska, S., Kamysz, W., & Kozlowski, H. (2011). *Dalton Transactions (Cambridge, England: 2003), 40*, 5604–5610.
Russell, T. L., Berardi, R. R., Barnett, J. L., Dermentzoglou, L. C., Jarvenpaa, K. M., Schmaltz, S. P., et al. (1993). *Pharmaceutical Research, 10*, 187–196.
Sachs, G., Meyer-Rosberg, K., Scott, D. R., & Melchers, K. (1996). *The Yale Journal of Biology and Medicine, 69*, 301–316.
Sachs, G., Weeks, D. L., Melchers, K., & Scott, D. R. (2003). *Annual Review of Physiology, 65*, 349–369.

Sanders, D. A., Gillece-Castro, B. L., Burlingame, A. L., & Koshland, D. E., Jr. (1992). *Journal of Bacteriology, 174*, 5117–5122.
Savarino, V., Mela, G. S., Zentilin, P., Mele, M. R., Lapertosa, L., Patetta, R., et al. (1996). *Gut, 39*, 508–512.
Scarlato, V., Delany, I., Spohn, G., & Beier, D. (2001). *International Journal of Medical Microbiology, 291*, 107–117.
Schär, J., Sickmann, A., & Beier, D. (2005). *Journal of Bacteriology, 187*, 3100–3109.
Schauer, K., Muller, C., Carrière, M., Labigne, A., Cavazza, C., & De Reuse, H. (2010). *Journal of Bacteriology, 192*, 1231–1237.
Schreiber, S., Konradt, M., Groll, C., Scheid, P., Hanauer, G., Werling, H.-O., et al. (2004). *Proceedings of the National Academy of Sciences of the United States of America, 101*, 5024–5029.
Scott, D., Weeks, D., Melchers, K., & Sachs, G. (1998). *Gut, 43*, S56–S60.
Scott, D. R., Marcus, E. A., Weeks, D. L., Lee, A., Melchers, K., & Sachs, G. (2000). *Infection and Immunity, 68*, 470–477.
Scott, D. R., Marcus, E. A., Weeks, D. L., & Sachs, G. (2002). *Gastroenterology, 123*, 187–195.
Scott, D. R., Marcus, E. A., Wen, Y., Oh, J., & Sachs, G. (2007). *Proceedings of the National Academy of Sciences of the United States of America, 104*, 7235–7240.
Scott, D. R., Marcus, E. A., Wen, Y., Singh, S., Feng, J., & Sachs, G. (2010). *Journal of Bacteriology, 192*, 94–103.
Seshadri, S., Benoit, S. L., & Maier, R. J. (2007). *Journal of Bacteriology, 189*, 4120–4126.
Sharp, P. M., & Li, W. H. (1987). *Nucleic Acids Research, 15*, 1281–1295.
Shirai, M., Fujinaga, R., Akada, J. K., & Nakazawa, T. (1999). *Gene, 239*, 351–359.
Siavoshi, F., Malekzadeh, R., Daneshmand, M., Smoot, D. T., & Ashktorab, H. (2004). *Helicobacter, 9*, 470.
Solnick, J. V., Hansen, L. M., Salama, N. R., Boonjakuakul, J. K., & Syvanen, M. (2004). *Proceedings of the National Academy of Sciences of the United States of America, 101*, 2106–2111.
Spronk, C., Żerko, S., Górka, M., Koźmiński, W., Bardiaux, B., Zambelli, B., et al. (2018). *Journal of Biological Inorganic Chemistry, 23*, 1309–1330.
Stähler, F. N., Odenbreit, S., Haas, R., Wilrich, J., Van Vliet, A. H., Kusters, J. G., et al. (2006). *Infection and Immunity, 74*, 3845–3852.
Stingl, K., Altendorf, K., & Bakker, E. P. (2002). *Trends in Microbiology, 10*, 70–74.
Stingl, K., & De Reuse, H. (2005). *International Journal of Medical Microbiology, 295*, 307–315.
Stingl, K., Schauer, K., Ecobichon, C., Labigne, A., Lenormand, P., Rousselle, J. C., et al. (2008). *Molecular & Cellular Proteomics, 7*, 2429–2441.
Stingl, K., Uhlemann, E.-M., Schmid, R., Altendorf, K., & Bakker, E. P. (2002). *Journal of Bacteriology, 184*, 3053–3060.
Stingl, K., Uhlemann Em, E. M., Deckers-Hebestreit, G., Schmid, R., Bakker, E. P., & Altendorf, K. (2001). *Infection and Immunity, 69*, 1178–1180.
Subramanian, S., Chaparala, S., Avali, V., & Ganapathiraju, M. K. (2016). *BMC Medical Genomics, 9*, 73.
Suerbaum, S., Thiberge, J. M., Kansau, I., Ferrero, R. L., & Labigne, A. (1994). *Molecular Microbiology, 14*, 959–974.
Szurmant, H., & Hoch, J. A. (2010). *Current Opinion in Microbiology, 13*, 190–197.
Takahashi, S., Igarashi, H., Nakamura, K., Masubuchi, N., Saito, S., Aoyagi, T., et al. (1993). *Japanese Journal of Clinical Medicine, 51*, 3149–3153.
Thomsen, L. L., Gavin, J. B., & Tasman-Jones, C. (1990). *Gut, 31*, 1230–1236.
Tomb, J. F., White, O., Kerlavage, A. R., Clayton, R. A., Sutton, G. G., Fleischmann, R. D., et al. (1997). *Nature, 388*, 539–547.

Tsuda, M., Karita, M., Morshed, M. G., Okita, K., & Nakazawa, T. (1994). *Infection and Immunity*, *62*, 3586–3589.
Valenzuela, M., Cerda, O., & Toledo, H. (2003). *Biological Research*, *36*, 429–436.
van Vliet, A. H., Ernst, F. D., & Kusters, J. G. (2004). *Trends in Microbiology*, *12*, 489–494.
van Vliet, A. H., Kuipers, E. J., Stoof, J., Poppelaars, S. W., & Kusters, J. G. (2004). *Infection and Immunity*, *72*, 766–773.
van Vliet, A. H., Kuipers, E. J., Waidner, B., Davies, B. J., de Vries, N., Penn, C. W., et al. (2001). *Infection and Immunity*, *69*, 4891–4897.
van Vliet, A. H., Poppelaars, S. W., Davies, B. J., Stoof, J., Bereswill, S., Kist, M., et al. (2002). *Infection and Immunity*, *70*, 2846–2852.
Vinella, D., Fischer, F., Vorontsov, E., Gallaud, J., Malosse, C., Michel, V., et al. (2015). *PLoS Pathogens*, *11*, e1005312.
Waidner, B, Melchers, K, Stähler, FN, Kist, M, & Bereswill, S. (2005). *Journal of Bacteriology*, *187*(13), 4683–4688.
Weeks, D. L., Eskandari, S., Scott, D. R., & Sachs, G. (2000). *Science*, *287*, 482–485.
Weeks, D. L., & Sachs, G. (2001). *Molecular Microbiology*, *40*, 1249–1259.
Wen, Y., Feng, J., Scott, D. R., Marcus, E. A., & Sachs, G. (2006). *Journal of Bacteriology*, *188*, 1750–1761.
Wen, Y., Feng, J., Scott, D. R., Marcus, E. A., & Sachs, G. (2007). *Journal of Bacteriology*, *189*, 2426–2434.
Wen, Y., Feng, J., Scott, D. R., Marcus, E. A., & Sachs, G. (2009). *Journal of Bacteriology*, *191*, 449–460.
Wen, Y., Marcus, E. A., Matrubutham, U., Gleeson, M. A., Scott, D. R., & Sachs, G. (2003). *Infection and Immunity*, *71*, 5921–5939.
Willett, J. W., & Kirby, J. R. (2012). *PLoS Genetics*, *8*, e1003084.
Williams, C. L., Preston, T., Hossack, M., Slater, C., & McColl, K. E. (1996). *FEMS Immunology and Medical Microbiology*, *13*, 87–94.
Wolfram, L., Haas, E., & Bauerfeind, P. (2006). *Journal of Bacteriology*, *188*, 1245–1250.
Wroblewski, L. E., Peek, R. M., Jr., & Wilson, K. T. (2010). *Clinical Microbiology Reviews*, *23*, 713–739.
Xia, W., Li, H., Sze, K. H., & Sun, H. (2009). *Journal of the American Chemical Society*, *131*, 10031–10040.
Xia, W., Li, H., Yang, X., Wong, K. B., & Sun, H. (2012). *The Journal of Biological Chemistry*, *287*, 6753–6763.
Xia, X. (2007). *Evolutionary Bioinformatics*, *3*, 53–58.
Xia, X. (2015). *Genetics*, *199*, 573–579.
Xia, X. (2018a). *Bioinformatics and the cell: modern computational approaches in genomics, proteomics and transcriptomics* (pp. 397–412). Cham, Switzerland: Springer.
Xia, X. (2018b). *Molecular Biology and Evolution*, *35*, 1550–1552.
Xia, X., & Palidwor, G. (2005). *The American Naturalist*, *166*, 776–784.
Yamaoka, Y., Kita, M., Kodama, T., Imamura, S., Ohno, T., Sawai, N., et al. (2002). *Gastroenterology*, *123*, 1992–2004.
Yang, X., Li, H., Cheng, T., Xia, W., Lai, Y. T., & Sun, H. (2014). *Metallomics: Integrated Biometal Science*, *6*, 1731–1736.
Yang, X., Li, H., Lai, T. P., & Sun, H. (2015). *The Journal of Biological Chemistry*, *290*, 12474–12485.
Zeng, Y. B., Yang, N., & Sun, H. (2011). *Chemistry*, *17*, 5852–5860.
Zeng, Y. B., Zhang, D. M., Li, H., & Sun, H. (2008). *Journal of Biological Inorganic Chemistry*, *13*, 1121–1131.

Kolibri 360 GmbH
Ferdinand-Jühlke-Straße 7
99095 Erfurt, Deutschland
produktsicherheit@kolibri360.de